Farm management and extension needs in Central and Eastern European countries under the EU milk quota

The following organisations have supported the CEEC Workshop and publication of this book:

- FAO Subregional Office for Central and Eastern Europe, Budapest, Hungary
- Ministry of Agriculture, Forestry and Food of the Republic of Slovenia
- Ministry of Education, Science and Sport of the Republic of Slovenia
- CR Delta, The Netherlands
- Semex Alliance, Canada
- Twinning project Netherlands-Slovenia: "Farming with Quota and Premiums", Twinning project SI04-AG-06

Farm management and extension needs in Central and Eastern European countries under the EU milk quota

Editors:
A. Kuipers
M. Klopčič
A. Svitojus

EAAP Technical Series No. 8

ISBN-10: 90-76998-92-2
ISBN-13: 978-90-76998-92-3
ISSN 1570-7318

First published, 2006

Wageningen Academic Publishers
The-Netherlands, 2006

Contents

Preface

The enlargement of the European Union with ten new countries has generated great expectations, and presents challenges and opportunities, particularly in the field of farm management. Among other prospects, farmers are expectant that agricultural production will become more profitable, that natural resources will be used in more effective ways, and that economic growth will increasingly stimulate commercialization of agricultural production. Without doubt integrated sustainable rural development programmes will be crucial in supporting higher living standards for the rural population. These transformations are linked with an urgent need to adopt new technically efficient and environmentally favourable technologies. Farmers will also need to reinforce their farm business and financial management capabilities. In order to achieve these goals, effective management and extension advisory services need to be developed in Central and Eastern European Countries, particularly for dairy producers, who have to effectively deal with the new phenomena of the EU milk quota system.

The EU is the biggest producer of dairy products in the world with a total milk production of about 145 million tonnes. The dairy sector is one of the most important areas in the EU agriculture and food domain and represents 14.1 percent of the value of its overall agricultural production.

Over the last decade the dairy sector in Central and Eastern European countries has undergone a transition from a centrally planned system to a market economy and radical changes have occurred in the milk production sector as dairy herds and milk production has declined. Following accession of the ten new Central and Eastern European states to the EU the framework of the milk production system changed significantly. The introduction of the Common Agricultural Policy (CAP) embodied the EU Milk Quota and intervention system for butter and skimmed milk powder, and was designed to stimulate growth in the scale of production, a decrease in the number of dairy farmers, increased herd size per farm, a movement of dairy production from less to more favourable regions, and a sharp decline in milk production costs through increasing yields and improved feeding. The implementation of the HACCP quality system, and EU and global sanitary and veterinary standards (Codex Alimentarius) necessitates adopting new high performance technologies assuring a high grade of quality and hygiene of milk as well as increasing productivity.

The transformation and adjustment of agricultural structures in the CEE countries under the CAP call for better managerial knowledge and decision-making skills for dairy farmers and extension service agents in the following areas: the milk quota system, planning and budgeting, marketing and in decision-making and problem-solving skills. The necessary adjustment in farmers' livelihood strategies and management practices requires innovation, improved access to information, farmer training and capacity building.

FAO's response to the challenges of the CEE countries and the emergence of transition economies is to provide technical assistance to its Member Nations through its Regular Programme based on the FAO Strategic Framework and the revolving Medium-Term Plan as well as through its Field Programme and by way of Technical Assistance. The FAO Regular Programme encompasses such activities as Farm Management Education in Central and Eastern European Countries, the Conference on CEE Agricultural Extension, Effective Management of Extension Advisory Services in CEE countries, Review of Farm Management in Extension Programmes in CEE countries and various stakeholder workshops.

FAO's Field Programme and Technical Assistance cover institutional aspects as a component of agricultural and rural development strategies as well as technical areas, particularly in Croatia, Bulgaria, Romania and Slovenia

The challenge facing CEE farmers is to make the necessary adjustments in their livelihood strategies and management practices, which calls for innovation, improved decision-making skills and better access to information, farmer training and capacity building. FAO's approach is to strengthen the capacity and functionality of extension services for sustainable delivery of farm management and agribusiness-related advice applying participatory and bottom-up approaches. Therefore the FAO Sub-regional Office supports many key activities related to agricultural extension which constitute new challenges for farmers from new EU countries. In this respect the CEEC workshop devoted to "Farm Management and Extension Needs in Central and Eastern European Countries under the EU Milk Quota" can provide strong support for Central and Eastern European Countries.

I would like to take this opportunity to express my appreciation to the principal organizers of this event, first of all to EAAP and the Government of Slovenia, and personally to Mr Abele Kuipers, Ms Marija Klopcic and Mr Arunas Svitojus for their highly professional approach to the organization of this workshop on "Farm Management and Extension Needs in Central and Eastern European countries under the EU Milk Quota."

Maria Kadlecikova
FAO Sub-regional Representation for Central and Eastern Europe

Introduction

In April 2004 the Central and Eastern European Countries entering the European Union (EU) had to implement among other regulations the quota system for milk. The introduction of a quota system is a very complicated process. It requires institution building, setting up administrative procedures, choices about the system, the choice of priority groups, the handling of the butterfat reference, control aspects, farm management and cost aspects, communication to farmers, etc. A quota system affects the dairy industry as a whole.

Exchange of experiences in this area between EU member countries and candidate countries would be very helpful, as was learned in 2002 and early 2003 in country to country and person to person discussions. Several Central and Eastern European countries visited Western European countries to learn from their experiences. More and more questions were raised. But because the questions from the various countries were quite similar, it was efficient to organise expert meetings for groups of countries together. However, we realised that no general forum existed between the candidate countries and the EU member countries about this topic. This led to the initiative to organise several meetings to support the introduction of the quota system and to establish a kind of platform. These initiatives are listed below.

Milk quota seminar in The Netherlands, May 2003

In May 2003, a milk quota seminar was held in The Netherlands. It concentrated on the administrative aspects of the introduction of the quota system. About 30 participants from the Central European countries attended this 2 ½ day workshop. A report was published named "Introduction aspects of milk quota systems in EU candidate countries". Organisers were Abele Kuipers, Gerrit Nijboer, Anneke Sellis and Gerard Krebbers from The Netherlands. The Summary of this workshop is included in this book as Appendix I.

Workshop in Budapest, December 2003

The workshop in December 2003 in Budapest about the milk quota system concentrated on the consequences of the quota and premium system for agriculture and especially for the dairy sector, and was at the same time a preparatory meeting for the Seminar to be held in Bled, Slovenia in September 2004. Some of the participants in Budapest also visited the meeting in The Netherlands. So links between the different meetings were established.

The meeting was organised by Arunas Svitojus, Chairman of the Working Group of Central and Eastern European Countries of EAAP and Abele Kuipers, Secretary of Cattle Commission of EAAP. The workshop was supported by FAO. The summary of this workshop is also included in this book as Appendix II.

Seminar in Bled, Slovenia, September 2004

The seminar in Bled, Slovenia on Saturday the 4th of September 2004 as satellite symposium of the EAAP (European Association for Animal Production) meeting had as title: "Farm management and extension needs in CEE countries under the restrictions of the EU milk quota". During this EAAP meeting, there was also a Round Table discussion about the enlargement of the EU with 10 new member States in 2004 (Title: "Enlargement of the European Union and Other Challenges for the European Livestock Production").

The seminar was organised by Abele Kuipers, Marija Klopčič and Arunas Svitojus. Mrs. Klopčič was secretary of the organising committee of EAAP 2004 in Slovenia.

A number of expertise contributions were presented from the West. All candidate countries, being Estonia, Latvia, Lithuania, Poland, Czech Republic, Slovakia, Hungary and Slovenia (except Malta and Cypress) presented a country report and seven other countries from Central and Eastern Europe (Romania, Bulgaria, Croatia, Albania, Turkey, Belarus and Georgia) told about developments in the dairy industry in their countries.

Questions to be addressed in the country reports were:
1. Which structural changes due to EU-quota system and premiums do you expect?
2. Do you have currently national support programs for the dairy sector?
3. What effect do you expect of milk quota on farm management?
4. How is the organisation of extension and of extension needs in your country?
5. What will be the biggest challenge for the dairy industry in years ahead in your country?

In Hungary and Czech Republic, production control was already introduced some years ago. So the representatives from Hungary and Czech Republic were expected to tell more extensively about their experiences with the quota system. This book contains a second contribution from Poland to present also the most recent data about this country. This exemption was made because of the significance of Poland as large dairy country.

The seminar in Bled in September 2004 was attended by more than 100 persons from 32 countries. Representatives from Ministries, Farmers Organisations, Industry, research institutes and extension and consultant services participated. All invited speakers and all country representatives with their country reports were present. It is pretty unique that there was not any no-show. This also expressed the interest the various countries had in the topic of this seminar. The performance of the speakers was excellent. This book contains all the contributions and country reports of the seminar.

I like to thank Marija Klopčič for all work done in organising this seminar and in the preparation of this book and Arunas Svitojus for his support.

Abele Kuipers

President to be of Cattle Commission of EAAP
Director Expertise Centre for Farm Management and Knowledge Transfer, The Netherlands

Overview papers

Expectations and challenges for farm management in Central and Eastern Europe and FAO response[1]

Stjepan Tanic[2]

FAO Subregional Office for Central and Eastern Europe, Benczur utca 34, 1068 Budapest, Hungary

Introduction

During the last decade and a half, agriculture in Central and Eastern Europe has been passing through the complex transformations of political and economic systems. Initially, all countries faced similar underlying challenges; however, because of substantial differences in the degree of prior market development and the level of political commitment, the progress in adopting and developing market-based agriculture in each country has differed substantially during the 1990s. Production fell, economies contracted and urban and rural poverty increased substantially. The process also implies re-creating and re-shaping of production and management in most countries, for which the re-engineering of market and support institutions is essential. The major difficulty has been to create the basic enabling conditions for successful private family farming.

Expectations and challenges

In the beginning of 1990s widely proclaimed and adopted ideas of democracy, benefits of market economy and accession to the European Union have created the expectation that vast natural potential will be used in a more effective way and agricultural production will be profitable and competitive in a relatively short time. Furthermore, there were also expectations that fast economic growth will increasingly cause commercialisation of agricultural production, while additional incomes from various forms of non-agricultural use of resources will offset the diseconomies of scale on a huge number of newly emerged small family farms and the surplus of labour in rural areas. On top of all, there was also a hope and expectation that integrated sustainable rural development programmes will be significant in supporting higher living standards for the rural population. However, the whole package still needs substantial time, improved institutional support and management capacity to be materialised.

Farmers in CEE countries have been faced with new political and economic environment that besides the structural changes also resulted with transfer of responsibilities and decision making to the farm level. The complexity of relationships between individual and public interest in agriculture and rural areas in particular, calls for an enlarged vision. The core remains farmers (of various kinds) managing plots of land on which decisions are made about what to produce, on what scale and where, when and how. Their decisions are affected by a

---

[1] This paper is based on the presentation given at the workshop "Farm Management and Extension Needs in CEE under the EU Milk Quota System", and as such has a limited scope. Its purpose is to briefly present circumstances and challenges in which farmers had to run their farms during the years of transition and transformation of agricultural systems in CEE countries, and to give a snapshot of FAOs response to the needs of farmers and governments to adjust to the rapidly changing environment.

[2] Farm Management and Agribusiness Officer, FAO Sub-regional Office for Central and Eastern Europe, Budapest, Hungary

commercial arena over which they may have some control (for example, through cooperatives or selling groups), or which may be entirely a private sector activity. Although privatisation measures have reduced the government arena by strengthening the commercial sector, government support, protection and guidance for agriculture through the use of economic measures and policy instruments is still largely interfering the decision making processes at the farm level. The public (consumers) also have wishes and beliefs about how food should be produced, and how land should be used. Furthermore, increasing commercialisation and integration into the food distribution chain is inhibited not only by the lack of market support institutions, but also by the lack of management capacity and knowledge.

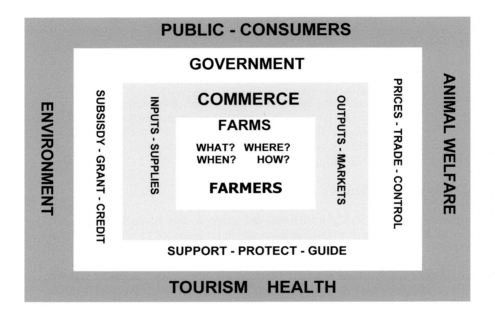

The challenge of transition and transformation made decision making process and management of farms more complex and demanding. Besides the need to adopt new technically efficient and environmentally neutral technologies farmers also need to enhance their farm business and financial management capacity. Farmers need to respond more rapidly and effectively to new market opportunities to minimise risks associated with commercialisation and globalisation.

The information and advice needed by farmers imposes new requirements on consultancy and advisory services for CEE agriculture, which have to be able to empower farmers to cope not only with the pressures originating from vertical coordination and concentration in food manufacturing and retail chains, but also from increasingly complex regulatory environment. In practice, studies have shown that farmers use multiple sources of information and advice, and construct their own systems from the sources which they find to be most reliable and useful.

There is a substantial variation in the methods of approach to farmers and in the users of services. These range from motivating producers to improve yields by using innovative technology; to working with farmers to understand the pressures they face in the local rural area, its environment and urban interests; to finding ways to satisfy consumer preferences;

and to exploiting opportunities for new sources of farm and off-farm incomes. What is common to the extension programmes is their contribution to responsible farm management decisions and plans for future farm viability; consultation to solve problems and help farmers to learn from experience.

FAO response

In response to the needs of FAO Member Countries in their efforts to improve managerial and entrepreneurial capacity of their rural dwellers, FAO Agricultural Management Marketing and Finance Service has been within the framework of Regular Programme implementing a number of components, alike:
- Enhancing Small Farmer Livelihoods,
- Agribusiness Development Targeted to Small and Medium Post-production Enterprises
- Enhancing Food Quality and Safety by Strengthening Handling, Processing and Marketing in the Food Chain and
- Agricultural Services - Data and Information Systems.

Objectives of those programmes are to:
- improve the support provided by public sector and civil society organisations to small farmers, including adjustments in their livelihood strategies, improved farm business management and income generation in the context of agricultural commercialisation and globalisation,
- increased capacity and efficiency of small and medium scale enterprises and entrepreneurs in Member Countries to offer consumers food and agricultural products through sustainable and profitable agribusiness ventures,
- to enhance food quality and safety during handling, processing, packaging, storage, transportation and marketing
- enhance capacity in countries for data collection, information access, and decision making in agriculture.

A number of assessment studies and stakeholder workshops[3] have identified that the major bottlenecks improved farm competitiveness, better use of labour, increased farm/household income and improved family farm livelihoods, are human resource capacity for adoption of new concepts and management tools, as well as the absence of policies addressing legal impediments to market development and creation of the favouring environment for emergence of new types of private smaller scale rural enterprises.

There is also a plenty of evidence of increasing farmer demand for help with farm management in CEE. The challenge has been to assemble and deliver information and advice to help producers to develop entrepreneurial skills, understand marketing, and become more efficient in the use of all their resources. The desired outcome is not only increased farm incomes but an efficient agricultural industry within a diversified rural economy.

Although extensive training has been provided both to management and field level staff of extension services in a number of CEECs, its impact and effectives, as identified by the *ex-post* evaluations of received assistance, has in a number of cases been limited. Due to horizontal dispersion and limited duration of training programmes for the staff of extension agencies, it did not induce creation of in-service capacity to sustain the acquired knowledge.

---

[3] Several studies and workshop proceedings are available at: http://www.fao.org/regional/SEUR/pubs_en.htm

Furthermore, farm business training materials lack sufficient comprehensiveness and quality that could assure their longer-term use by the trainers and the rest of the staff.

In such cases FAO is through the technical assistance projects contributing to improved market access and increased farm income for small and medium-sized farmers by strengthening the capacity and functionality of extension services and farm data systems for sustainable delivery of information, farm management and agribusiness related advice suited to the needs and based on the demand of small and medium-sized commercially oriented farmers and their organizations. The approach is to improve the potential of small- and medium-sized farmers, producer groups and local private-sector buyers to market access and leverage in the food chain, by facilitating establishment of sustainable farm-agribusiness linkages.

Contribution to national capacity building is achieved through a close working relationship between local staff, national and international consultants, and FAO specialists. This is accomplished through intensive training-courses for trainers themselves in: farm management, marketing, farm data and market information analysis and use, group presentation and communication skills and farm-agribusiness development. The approach also comprises of provision comprehensive – tailor made training materials in the above-mentioned areas to help in building up a critical mass of capacity for selected staff of advisory agencies who would then be able to sustain and further develop acquired skills and techniques. Finally, the whole knowledge transfer process is exercised through a series of field trainings in farm business management, marketing and the practical establishment of farm-agribusiness linkages at selected pilot sites and in interaction with farmer groups and private-sector buyers.

Conclusion

Small individual farmers throughout CEE countries in particular are facing unprecedented opportunities and risks as a result of privatisation and restructuring of agricultural sectors including market liberalization, privatization of agricultural services, trade globalization and integration of food chains. The challenge facing small farmers is to make necessary adjustments in their livelihood strategies and management practices, requiring innovation, and improved decision making.

In response to the needs of FAO Member Countries in their efforts to improve managerial and entrepreneurial capacity of their rural dwellers, FAO is implementing a number of regular and field programmes and projects. FAO approach is to contribute to improved market access and increased farm income for small and medium-sized farmers by strengthening the capacity and functionality of extension and information services for sustainable delivery of farm management, marketing and agribusiness related advice suited to the needs and based on the demand of small and medium-sized commercially oriented farmers. It is focused on the improvement of the capacity of farmers, producer groups and local private-sector buyers, to establish sustainable farm-agribusiness linkages by adoption of participatory and bottom-up approaches.

References

FAO. 1999a. Central and Eastern European Workshop on Needs and Potentials for Farm and Farming Systems Data. In: Proceedings of the Workshop, Budapest, Hungary, 15-19 December, 1997, REU Technical Series 60, Rome, Italy.
FAO. 2000. Market Information Services in Central and Eastern European Countries. In: Report of FAO/GTZ Workshop, Budapest, Hungary, 14-17 May 2000, Rome, Italy.

FAO. 2001. The impact of structural adjustment programmes on family farms in Central and Eastern Europe. Proceedings of FAO expert consultation, Budapest, Hungary, 20-23 January 2000, http://www.fao.org/regional/seur/pubs/.

Rolls, M. J. 2001. Review of Farm Management in Extension Programmes in Central and Eastern European Countries. Working paper, FAO, Rome, Italy.

Tanic, S. 2001. Individual Farms in CEE: From Structural Adjustment to Family Farm Livelihoods. FAO Workshop on Individual Farms in Eastern Europe and Commonwealth of Independent States – In: Proceedings, September 28-29, Budapest, Hungary, FAO, Rome, Italy.

Tanic, S. 2002. Traditions and needs for farm data in the main types of farms in Central and Eastern Europe, FAO Workshop "Sustainable Farm Household Information Systems for Improved Livelihoods and Reduced Hunger and Poverty – In: Proceedings, 4-7, December 2001, Rome, Italy.

Tanic, S. & Dixon, J. 2002. Farming Systems Based Strategies for Improved Rural Livelihoods in Eastern Europe and Central Asia, 17th Symposium of the International Farming Systems Association, November 17 - 20, 2002, Lake Buena Vista, Florida, USA.

Tanic, S. & T. Lonc, 2003. Farm Commercialisation and Income Diversification on the Road to EU Accession. In: Proceedings of FAO Workshop, Prague, Czech Republic, 2–6 November 2003, FAO, Rome, Italy.

Overview and characteristics of quota systems

Abele Kuipers

Expertise Centre for Farm Management and Knowledge Transfer, Wageningen University and Research Centre, De Leeuwenborch, Hollandseweg 1, 6706 KN Wageningen, The Netherlands

1. Introduction

The introduction of a quota system is a very complicated process. It requires institution building, setting up administrative procedures, choices about the system, the choice of priority groups, the handling of the butterfat reference, control aspects, farm management and cost aspects, communication to farmers, etc. In fact the quota system affects the dairy industry as a whole (Figure1).

This contribution covers some aspects of the following topics:
- EU agricultural Policy
- Choices to be made by introduction of quota system
- Farm Management under quota
- Communication about quota system

Figure 1. Dairy farming under quota conditions is not the same as before.

2. EU Agricultural Policy

EU Commissioner Franz Fischler presented in 2002 a plan for the common agricultural policy. General objectives of this plan are:
- To improve competitiveness of EU agriculture
- To promote a more market oriented and sustainable agriculture
- To put more emphasis on rural development

The dairy proposal contains the next policy measures:
- Milk quota stay till year 2015
- Product price support changes to income support (direct payments)
- Intervention price is expected to decrease by about 20%.
- Income support 2004 - 2007 linked to kg's milk quota. The intention is that the income support is nearly equal to the reduction in price. But will this be realised?!
- Income support will be in 2007 decoupled from kg's

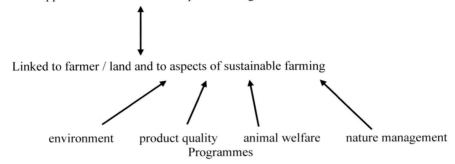

The national quota for all EU member states, including the new countries, are illustrated in Figure 2. It is quite clear that Poland as new EU-country belongs to the large dairy countries in Europe, ranking 6th in quota amount of the 25 EU-countries.

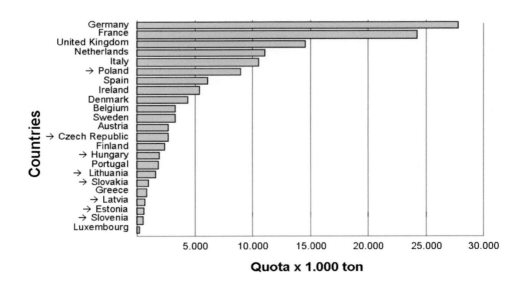

Figure 2. Quota assigned to EU-member states ($\rightarrow$ indicates the new EU-states; Malta and Cyprus not listed).

3. Choices to be made by introduction of quota system

The implementation of the quota system allows different options. The choice of options is very important for the development possibilities of the dairy sector and the individual farm. Especially the way quota transfer is arranged, the built-up of a national reserve and the division of the country in regions or not is essential in this context. The various choices to be made are discussed below.

3.1 Reference year

The reference year is the year on which the individual quota allocation to farmers is based. For the new member states this is usually 2002, 2003 or 2004 or a combination of these years. Each member state has been assigned a national quota by the EU (see Figure 2). The national quota amount is distributed over the national reserve and the individual producers. In some states the national quota is first divided between regions and/or between the milk purchasers and in a second step assigned to the individual producers by regional authorities and/or by the purchasers of milk (cooperatives or processing plants). The national quota and individual quota are affected by the development of the cow population and milk volume in the past. The decrease in the number of cattle and corresponding production volume in the Central and Eastern European countries is illustrated in Figure 3.

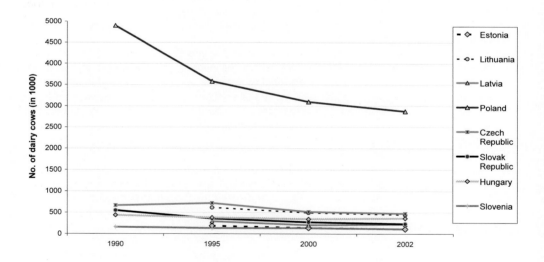

Figure 3. Development of number of cows in new EU-member states.

3.2 National reserve

A national reserve is needed to help farmers in specific problem situations and / or to stimulate certain national policies on structural development. The national reserve is meant for:

a) **To give to problem farmers at start**
Some possible problem groups:
- Farmers who invested in cow housing
- Starters
- Farmers' families with tragic circumstances
- Herds with diseases

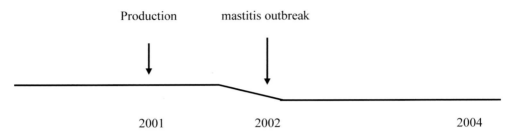

Experience shows that there is a lot of interest for additional quota. As example, in The Netherlands 50% of the farmers requested more quota in the initial period of the introduction of the quota system. We advise to limit the distribution of additional quota to very specific and well described groups of farmers. And we recommend to install a neutral independent committee to judge the requests for additional quota. A national reserve of about 3% of the national assigned quota should be enough to serve for this purpose.

b) **To provide for structural changes**
It is possible to have a more liberal or a more social approach to distributing quota. With a liberal approach the quota transfers are left to the market. With a social approach, the national authorities collect quota at national or regional level and have guidelines for the distribution of this quota. In this case the quota system is used as a development tool for the dairy sector and for rural development. As possible conditions for providing this additional quota from national reserve can be mentioned:
- farmers with development plans
- regions with land reform plans

Two questions determine largely the success or failure of such a social approach: who's in charge and who distributes this quota? In this context it is crucial to have a committee or agency which is independent, that is in charge of distributing the additional quota.

3.3 Quota transfer

There exist various ways of organising the transfer of quota from one farm to the other. The system to be chosen can be decided by the national authority. Free quota emerges when a farmer stops his farm or changes to another branch of income. The national authority can decide to block individual transfers of quota and instead collect all free quota centrally. It can also decide to collect part of the quota transfers (for instance a certain percentage of each quota transaction). With this policy farmers will receive quota from the national reserve in one way or the other. Quota can also be distributed by a Quota Exchange Bureau, by free transfer of quota and by transfers in which the quota is attached to the land or to the farm as a whole. The various routes will be described.

A. Quota Exchange Bureau

With a Quota Exchange Bureau quota transfer is organised centrally as depicted in the scheme below and illustrated in the photo of a stock exchange market (Figure 4):

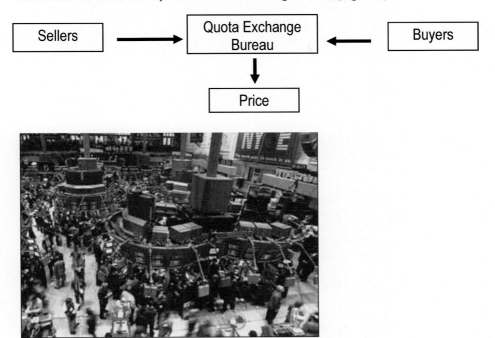

Figure 4. Stock exchange as illustration of exchange of assets, like in a Quota Exchange Bureau.

With this system quota becomes a capital asset. For each round of quota exchange, one price exists. This price is the same for all sellers and buyers who participate in the market. Denmark and Germany have Quota Exchange Bureaus. In Denmark onenational Quota Exchange Bureau functions. In Germany each region has it's own exchange and therefore it's own price. Also in Ireland each region has its own quota exchange. The Farmers Cooperatives most times act as the purchasers of the milk. These cooperatives execute the transfers of quota. However, one price is used for all regions together. This quota price is an artificial price, set by a national commission on an annual basis. Initially also Sweden did have a Quota Exchange Bureau system. But some years ago they changed to a free transfer system of quota.

B. Free transfer of quota

When free transfer of quota exist, sellers and buyers discuss prices for each transaction.

Seller ⟶ Buyer

In this system quota also becomes a capital asset. The national authority often sets some particular conditions for transfers. For instance transfers can only be done in a certain period of the year. Sometimes a certain percentage of the transferable quota is taken away and donated to the national reserve.

The system of free transferable quota is exercised in The Netherlands, UK and Sweden (last couple of years). The system has advantages and disadvantages as described in the scheme below:

Advantage
- System is flexibel

Disadvantage
- System of transfer is not transparent
- Intermediaries influence the market
- Because of this, probably higher quota prices

The free transfer system leads still to-day to a lot of discussion in some Western European countries. For instance, the level of quota prices is one of the main concerns of the dairy community in The Netherlands. Farmers' organisations believe that the intermediaries (firms who are marketing quota, similar like trading of cattle and stocks) are artificially elevating the quota prices. Costs of buying quota have become a major cost factor at farm level.

C. Transfer of quota linked to land

In France quota is linked to the land or to the farm enterprise. This was initially also the case in The Netherlands. This implies that quota can only be transferred with a certain acreage or with he whole farm. With this system quota cannot clearly be distinguished from other capital assets. The capital asset exists of land + quota. Off course land with quota is more expensive than land without quota. But it depends on the circumstances how prices develop.

As example: take a farmer with 120.000 kg of quota and 12 ha's of land. On average he has 10.000 kg of quota resting on one ha. When selling 20.000 kg of quota, he has to sell also 2 ha's of land. Often regulations permit that the new owner of the quota rents 2 ha's of land in stead of having to buy this land..

Some advantages and disadvantages of this system are mentioned in the scheme below:

Advantage:
- Quota is integral part of farm
- There exists not a clear separate price for quota

Disadvantage:
- System is less flexibel

D. Leasing

The tool of leasing means that quota is temporary used by somebody else than the owner/permanent user of the quota. Usually a period of 1 year is indicated. Contracts have to be renewed each year.

Lease prices differ from country to country in a similar way as the quota prices do. In the Netherlands lease prices were around euro 0.18 per kg milk in 2004, while the milk price at the farm gate was about euro 0.35 per kg milk. For comparison, quota prices were about euro 1.40 per kg milk.

The experience is that the persons/companies/intermediaries that lease quota are often not active in agriculture anymore or want to stop farming gradually. So money flows out of the sector in hands of middlemen, etc. Many older farmers continue farming while renting their quota on a yearly base to professional farmers. The possibility of leasing is slacking down the number of older farmers or small farms leaving the sector. This affects a restructuring of the sector in a negative way.

Problems with quota transfer

The price of quota becomes a cost factor at farm level. An impression of prices paid per kg of milk quota bought in several countries is presented in the box below. In each country a certain fat percentage is taken as base.

Price per kg of milk quota – 2004/2005	
Netherlands	4 – 6x milk price
Germany	0.7 – 2.5x
Denmark	0.9 – 1.5x
UK	0.8x
France	0.4x

In The Netherlands the price of quota is expressed per kg of fat rather than per kg of milk. High quota prices in this country have led to the sayings:
• Farmers who quit farming become rich (they sell their quota and farm)
• Farmers who start farming have higher costs (they have to buy quota as an additional asset)

However, it is planned that the new EU policy of direct payments instead of price support will result in lower milk prices. This may effect the quota prices in a downward way.

Another question that emerge with quota as capital goods: Who is the owner of quota? Is this the owner of the land and /or farm or is this the renter of the land.

In the box below the number of quota transactions in one year is given for The Netherlands. It illustrates the quick turnover of quota after the system was in use for 20 years in this country.

Average transfers of quota/year:		
• With farms	1.000 transactions	400 mln kg
• With or without land	15.000 transactions	350 mln kg
• Leasing	25.000 transactions	650 mln kg
Per year: in total 13% of national quota		

Advice about quota transfer

Different policies exist concerning quota transfer. The reality is that each country adapts a system that is more or less fitted to the own situation. What can we learn from the systems practised in the Western -European countries.

We learn that free quota transfer gives flexibility. Structural developments are not blocked. But when quota prices increase, the cost level of farms also increases. This is especially bothersome when the farm goes from one person to the other, for instance from father to son. Then the start value of the farm enterprise is higher than without quota.

In general it is advantages to keep quota price low or non existent. The prices in France have always been relatively low. Therefore the system in France may be interesting to study. France has regions and quota is linked to the land or to the whole farm. In this way quota is not clearly a separate capital asset.

Table 1. Size of farms in some countries (2004).

Country	farms	cow/farm	milk prod/cow
Netherlands	30,000	52	7,500
Lithuania	130,000	2.5	5,015
Poland	450,000	4.5	3,840
Hungary	30,000	12	6,317
Estonia	7,120	16	5,119
Czech Republic	3,400	212	5,718

The ideal system of quota transfer may also differ with herd size. Table 1 shows that herd sizes differ significantly between countries. When herd size is small it may be expected that many small farms will stop milking, c.q. are not allowed to deliver milk to the dairy plants. This is caused by milk quality and economic reasons. For the dairy sector it is the best that this quota goes for no price to other farmers. That will keep investments in the sector limited and cost levels competitive. Quota transfer without money transactions are only possible when free transfer is not allowed. In this case all free quota flows to the national reserve. The government has to distribute this quota among the farmers which stay. Also whole farm transfer, like in France may be an option for dairy countries with a small farm structure.

Leasing is an arbitrary tool to be used in practice. Because of the disadvantages listed in the box below, we advise **not** to allow the tool of **leasing** as part of the national regulation.

Disadvantage:
- Money flows out of sector
- Limits structural development

3.4 Regions or not

The philosophy behind the establishment of regions is that quota is forced to stay within each region. It protects certain regions from loosing quota. This is especially the case for the less favoured agricultural regions.

In the box below countries are mentioned that have installed quota regions.

- Germany   21 regions   quota price: 0.20 – 0.90 € / kg milk
- France   90 regions
- Poland   16 provinces
- Ireland   15-20 regions one quota price: In 2004 0.31 € and in 2006 0.13 €

Each region has it's own characteristics. As example, the quota prices of the various regions in Germany in 2004 varied from Euro 0.20 - 0.90. The lower prices exists in the former East-Germany regions and the highest prices in Southern Germany (Bayern). A first indication of quota price in Poland is 0.20 € per kg and in Slovenia 0.15 − 0.30 € per kg. But such a price may change considerably in years to come. Also number of quota transfers in the various regions may differ significantly, indicating that more or less restructuring of the dairy sector is going on in different parts of a country.

The advantages and disadvantages of instalment of regions are summarised in the box below:

Advantage:
- protection of certain regions to maintain dairy husbandry in those areas

Disadvantage:
- economic developments restricted to the borders of a region

---

3. 5 Dairy factory or national quota bureau

The administration of the quota system can be established at the purchasers' level and/or regional level and / or at the national level. When the administration is at the purchasers level, often dairy factories or co-operatives take care of (part) of the administration. At national or regional level usually an agency (agencies) is (are) established to perform the administrative duties, most times in combination with the administration of the EU- agricultural premiums.

The administration can be spread over many purchasers as is shown in Table 2. In a relatively small dairy country like Slovenia already 98 purchasers are functioning, while this number is 280 in the largest dairy country Germany. A question remains how efficient the split up of (part of) the quota administration over so many units will be.

Table 2. The number of purchasers in different countries (2004).

Country	No. of purchasers
The Netherlands	50
Germany	280
Slovenia	98
Hungary	65

Even more complicated is the administration of direct sales. Direct sales are often not well administered. So some estimation work is part of the job. Direct sales vary strongly over the different countries as illustrated in Table 3. The central or regional agency takes care of the administrative work.

Table 3. Direct sales as part of total milk sales in different countries (2004).

Country	Part of direct sales
The Netherlands	1%
Germany	2%
Slovakia	2%
Poland	5%
Slovenia	17%
Austria	20%
Lithuania	24%
Latvia	33%

Figure 5. Explanation to farmers and experts from organisations about quota systems.

Farm management, extension and quota

Before the introduction of the quota system an increase of income from the activities on the farm could be realised by expanding the farm by means of more livestock and inputs, like feed and fertilisers. The annual costs of the needed investments and inputs can be compared with the estimated annual revenues and based on that information an adequate decision can be made about the future farm set-up. As a very simple example of this: the additional revenue of 1000 kg more milk (the marginal revenue) is compared with the additional cost of feeding more concentrates to the cows to achieve this (the marginal cost). However with the introduction of the milk quota system this formula is not working any more.

As we know, the Central and Eastern countries introduced national quota in 2004. The farmers were assigned individual quota. It appears that in most countries at the start considerable amounts of quota are available in the national reserve. Under certain conditions farmers can apply for additional quota from the national reserve which gives them the opportunity to expand their amount of quota without costs. The quota that is left in the national reserve functions as a kind of buffer. When farmers exceed their individual quota this buffer is used to compensate the surplus. For the first couple of years to come this gives the farms flexibility in production and ,in fact, allows them to exceed their individual quota. However in the longer run the resources (available quota) from the national reserve may become limited or even exhausted. For instance, the average production per cow is expected to increase with 2 – 3% per year under the conditions of good management and the utilisation of well defined breeding programs. This implies that without any interference this eventually leads to a situation of overproduction at national level. For comparison: the majority of the Western European countries that implemented the quota system in 1984 did reach this status and those countries pay (paid) a large amount of penalties to the EU.

For the individual farmer in such a situation this means that he needs to review his farm strategy and adjust his management to obtain good economic results. Moreover, in the coming years the farmer needs to adjust his production volume to the quota available. Overproduction at the end of a quota year should be prevented to avoid penalties. For this several management options can be thought of varying from actions for the short run that enable the management to be adapted quickly during the current quota year till options for the longer term, that take more time to adapt i.e. result in structural changes in the farm set-up.

Some examples of options for farmers for the short and longer run to adapt the management in a situation when the milk production is expected to exceed the assigned quota:

- Selling low productive cows - this will lead to an immediate reduction in the milk volume and can be considered a good option for the short run. It consequently increases the average production per cow at the farm.
- Feeding the excessive milk to calves - the amount of milk that can be given to the calves is limited but it can be of some help in not exceeding the quota, especially at small farms.
- Feeding less concentrates - in this way the maximum production capacity of cows is not used and the cows will produce less milk. On the contrary savings on feeding costs are realised.
- Using less nitrogen - the quantity and perhaps quality of hay or silage will reduce which may results in a lower milk production of the herd.
- Increasing the young stock or number of beef animals - this option can be used to balance the feed situation at farm level when a reduced number of dairy cows are needed to fulfil the quota. The extra land can off course also be used for sheep or crops, etc.
- Increasing production per cow with alternative use of land - this implies indirect a reduction of the number of cows. The increase of production can be achieved by different breeding strategies but also by improvement of management. Influencing the genetic source of the animals is often a long path which becomes effective in 5 to 10 years time and is typical an option for the longer run. The not utilised land is used for other animal species or another form of diversification in the farm set-up.
- Obtaining quota - this option looks at the future. Exceeding the quota can be avoided by increasing the quota amount, but this strategy is also chosen to maintain or increase the size of the herd.

When milk quota was introduced in the Western European countries several solutions were found to deal with overproduction at farm level. For instance, in The Netherlands a lot of farmers initially fed excessive milk to the calves in stead of using milk powder. Farmers also decided to sell the less productive cows. In the United Kingdom a reduction in concentrate feeding was practised which might have been related to the relatively high price of the concentrates compared to for instance the situation in The Netherlands. In other words, solutions depend on the circumstances.

Under the restriction of farming with quota and premiums also animal breeding can be subject to adaptation. With the introduction of quota the economic value for the selection trait of quantity of milk is expected to decrease substantially, whereas the economic value for the selection trait of protein will stay more or less the same (protein content is not part of the quota system). For this reason within the breeding programs in general more emphasis has been put on protein content, while secondary traits as mastitis, fertility, longevity and beef traits receive more emphasis with the final goal of achieving better economical farm results.

For some farmers all of these options might not be possible or not challenging enough. Especially when specialisation in dairy farming is not a feasible option because of certain limitations in land availability, etc., then diversification of the farm set-up may be an option. For instance, some countries or regions have good possibilities in the area of agro-tourism and cottage industry. Furthermore, some countries have a long tradition of producing special regional products and some farmers might be able to start supplementing their income by producing these special products. But of course, also the inclusion of other animal species in the farm set-up will be considered. In conclusion: it's wise to discuss strategies for the future to take the future in own hands.

The impact of the changes resulting from the introduction of quota and premiums must not be underestimated and it is therefore important that not only farmers, but also supporting institutions meet regularly for a longer time span, i.e. several years to be updated on the regulations and especially on available management options to deal with it. In this way farmers and advisers will be equipped with knowledge and tools to anticipate actively on the future and make decisions for the best set-up of the individual farm and the development possibilities for the farming community as a whole.

Developments in the milk market in the enlarged European Union

Clemens Fuchs

University of Applied Sciences Neubrandenburg, Postfach 110121, 17041 Neubrandenburg, Germany

Introduction

In May 2004 the EU increased to 25 members. The integration of the 10 new member states is not only a political task but also an economical challenge. The enlargement from the EU-15 to the EU-25 changed the framework for milk production significantly. The introduction of the Common Agricultural Policy (CAP) in the new member states includes the intervention system for butter and skimmed-milk powder (SMP) as well as the restriction of the milk production due to the quota system. In the common market of the EU-25 all milk producers compete with each other. Regions with enormous disadvantages in farm structure face a major change to become competitive. The larger market with additional 75 Million consumers, which is about one sixth of the total EU-25 population, offers bigger marketing prospects for dairies (Table 1). But not only the number of consumers has been enlarged, also their income might increase. To anticipate the most important point an increasing income and adopted consumer habits will lead to more demand for fresh milk products and cheese. Due to the enlargement in particular the milk market in the EU has grown by one fifth with the enlargement. This is a significant change for one of the most important agricultural sectors in the EU.

Table 1. Main characteristics.

	EU-15	+	N-10	= EU-25	
Area	3.154 (81%)	+	737 (19%)	= 3.891	000 km^2
Population	379,6 (84%)	+	74,8 (16%)	= 454,4	Mio.
Milk production	121,9 (85%)	+	22,1 (15%)	= 144,0	Mio. t
Milk producer	734 (32%)	+	1.589 (68%	= 2.323	1.000
Herd size	29		3	15	cows
Milk yields	6.050		4.050	5.730	kg

Milk is an important sector

Milk is about one third of the value of animal production and one seventh of the value of the overall agricultural production in the EU-25. In Table 2 the countries are ranked according to their share of milk production in relation to the value of total agricultural production in 2002. With a production share of about 25% milk is most important in Northern Europe (Finland, Sweden, Ireland), followed by the Baltic states (Estonia, Latvia and Lithuania) with a share of

20 to 23%. Also in the Czech Republic, in Slovenia, Slovakia and in Poland the milk production has a value between 15 and 20%, which is comparable with countries like Germany, Denmark, the Netherlands, Austria or the United Kingdom. Contrary are the Southern European countries, where other products like fruits and vegetables, wine and olives have a higher importance. Here milk production looses emphasis and at the same time the demand for milk import increases.

The process of milk production is labour-intensive and is often located in less-favoured areas. Therefore changes in the milk market have mostly implications on the employment in rural areas and on the income development in the farming sector.

Changes which are forced by political decisions (introduction of quotas, investment incentives, increase of market subsidies or the reverse) and which increased competition by opening the boarders force especially farmers with small scale production structures to undertake huge efforts to become competitive.

Table 2. The value of agricultural production and the share of animal and milk production in the EU member states in 2002.

EU-member state	Value of total agricul. production in Mio EUR (2002)	Ag. prod. in % of EU-25	Animal prod. in % of ag. prod.	**Milk in % of ag. prod.**	Milk in % of animal prod.
Luxembourg	249,6	0,1%	61,4	**34,9**	56,8%
Finland	4.092,7	1,3%	48,1	**26,5**	55,1%
Ireland	5.745,6	1,9%	70,1	**24,4**	34,8%
Sweden	4.513,2	1,5%	48,5	**23,8**	49,2%
Estonia	**400,1**	**0,1%**	**53,6**	**23,2**	**43,2%**
Latvia	**531,4**	**0,2%**	**46,4**	**22,0**	**47,3%**
Germany	41.329,4	13,4%	46,1	**21,0**	45,6%
Lithuania	**1.049,2**	**0,3%**	**42,5**	**19,8**	**46,5%**
Czech Republic	**3.281,3**	**1,1%**	**48,7**	**19,6**	**40,2%**
Denmark	8.341,9	2,7%	54,3	**18,3**	33,7%
Netherlands	20.021,2	6,5%	39,5	**18,1**	45,9%
Slovenia	**1.062,2**	**0,3%**	**48,6**	**17,2**	**35,3%**
Austria	5.301,8	1,7%	47,3	**16,7**	35,3%
United Kingdom	23.494,4	7,6%	55,9	**16,3**	29,1%
Slovakia	**1.506,6**	**0,5%**	**50,5**	**15,4**	**30,6%**
Poland	**13.058,6**	**4,2%**	**48,9**	**14,8**	**30,3%**
Belgium	7.012,9	2,3%	50,9	**12,7**	24,8%
France	63.265,7	20,6%	36,9	**12,4**	33,5%
Malta	**140,1**	**0,0%**	**45,2**	**12,3**	**27,1%**
Portugal	6.257,7	2,0%	38,1	**12,1**	31,9%
Italy	42.870,9	13,9%	31,1	**10,3**	33,2%
Hungary	**5.890,3**	**1,9%**	**43,4**	**10,2**	**23,6%**
Greek	11.708,8	3,8%	23,5	**8,3**	35,4%
Spain	36.240,5	11,8%	34,1	**6,3**	18,4%
Cyprus	-	-	-	**-**	-
N-10	26.919,7	8,8%	47,5	**15,0**	31,5%
EU-15	280.446,2	91,2%	40,4	**14,0**	34,8%
EU-25	307.365,9	100,0%	41,0	**14,1**	34,4%

Source: EU-Commission: The 2003 Agricultural Year, Tab. 3.1.1, own calculations.

Current milk production and future milk quotas

Despite all changes due to advanced production techniques in breeding and in changes farm structure the milk quota has proofed to be the stabilising element in the EU milk market in the past. The quota was introduced in 1984 and was prolonged in the latest CAP-Reform up to 2014/15. During the accession negotiations the new member states and the EU agreed on a quota of 253 kg per capita on average for the N-10, while the quota in the EU-15 is at 316 kg per capita (see Appendix).

One could expect that the quota system would protect the milk market further on, but due to the CAP-Reforms as well as due to the EU-Enlargement major changes can be recognised. These changes are the adoption of the current production level to the political agreed quota amount in the new member states. Other changes are primary the reduction of intervention prices for skimmed-milk powder and for butter, which effects also the producer milk prices. The gap between current milk production (2002) and future milk quotas (2008/09 to 2014/15) is biggest in Poland, where a reduction of 24% would be necessary (Figure 1). A potential overproduction is also recorded for Slovakia (15%), Latvia and Slovenia (each 14%), Hungary (13%) and Lithuania (7%). These six countries would have to reduce cow stocks significantly more than only the compensation of increasing milk yields.

In 2002 about 95% of the cow's milk production in the EU-15 was delivered to dairies. In the new member states this proportion only amounts 69%, while the lowest values have been recorded in Latvia (53%), Poland (61%), Lithuania (63%), Slovenia (75%) and Hungary (77%). In all other countries this measure is above 80%. A relative low delivery percentage to dairies goes along with a high level of on-farm consumption and direct marketing of milk. Especially countries with small structured farms belong to this category.

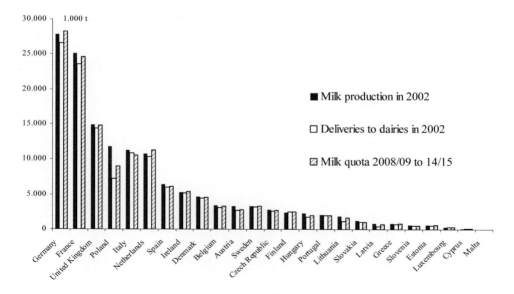

Source: EU-Commission Accession proposal 27.10.2003 en.pdf; ZMP Agrarmärkte in Zahlen 2004 Milch, p. 65

Figure 1. Milk production and deliveries to dairies (2002) and milk quotas (2008/09 to 2014/15) in the EU-25 countries.

Production structure

From the 145 Mio. tonnes of milk production in the EU-25 the new member countries hold a share of 22 Mio. tonnes or 15 percent. Because of the low average milk yields per cow and year, the proportion of cow stock is higher with 20%. While some of the new members have large cow herds (Czech Republic and Slovakia with an average of 140 heads), others have very small structured dairy herds with only one, two or three cows per farm (Poland, Lithuania, Latvia, see Figure 2). The milk production of the small units is mostly for own consumption or direct marketing. Because of these numerous but small farms the majority of the milk producers belong to the new member states with 68% or two-third of all milk producers in the EU-25. In absolute numbers there are 1.6 Million dairy farmers in the new member states and "only" 0.7 Million in the old member states, all together 2.3 Million in the EU-15. The countries with the most milk producers are Poland (1.2 Million), Lithuania (0.225 Million) and Latvia (0.075 Million).

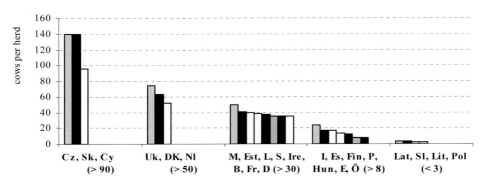

Source: http://europa.eu.int/comm/agriculture/agrista/2003/table_en/index.htm (Table 3.5.3.6); Wohlfarth (2004)

Figure 2. Average size of cow herds (EU-15 in 2001 and CC-10 in 2000).

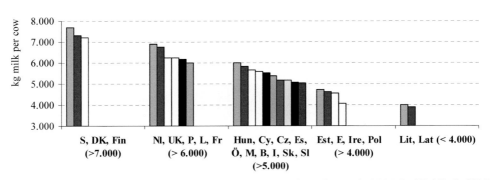

Source: http://europa.eu.int/comm/agriculture/agrista/2003/table_en/index.htm (Tab. 4.20.0.1); (Wohlfarth, 2004).

Figure 3. Milk yields in the EU-25 countries.

In conformity with the structure of milk production the yield level in the new member states is very different. In Poland the average size was 2 cows per holding (2000) with a yield of 3,774 kg per cow (2001) (Wohlfarth, 2004) while in the Czech Republic on average 140 cows per herd were kept with a milk yield of 5,400 kg (Figure 3). In comparison: The average size of a dairy farm in the EU-15 was 29.4 cows and the milk yield reached 6,000 kg in 2001.

Summarising the structural situation one can say that in the new member states currently a share of 15% of the EU-milk is produced by 68% of the dairy operations, while in the old member states 85% of the EU-milk is produced by only 32% of the holdings. The future development of the yields will influence the necessary number of cows to produce the quota. The yields are expected to increase up to 7,000 kg in the EU-15 and up to 5,200 kg in the N-10 by the end of the decade. In their latest prospects for the agricultural markets in the EU-25, the EU-Commission forecasts the stock of cows to decline from 24.5 Mio. in 2002 to 21.6 Mio. in 2010, while 17.4 Mio. cows will be kept in the EU-15 countries and 4.1 Mio. in the N-10 countries (Figure 4).

For countries with dominating small farm structures, an enormous structural change is expected. This necessary change can be overcome the sooner the general economy, employment and income are growing. The hygienic regulations of the EU will accelerate this process, because investments to improve the product quality will be profitable only surpassing certain minimum scales. After the interim regulations have been passed in 2006, for many semi-subsistence farms the delivery of milk to dairies has to be stopped. To overtake production capacities from smaller farms, there have to be favourable economic conditions for growing farmers. The national state and its administration can support this development, for example in the kind of regulation for quota transfer between farmers.

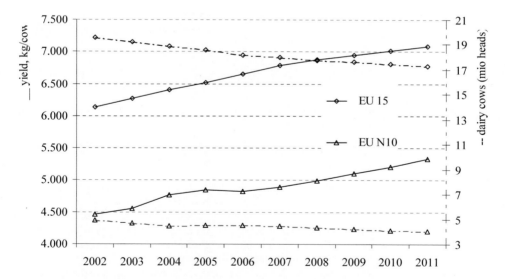

Source: EU, Prospects for Agricultural Markets, Brussels July 2004.

Figure 4. Prospects of the development of the number of cows and the milk yields in the EU-15 and in the N-10.

Administrative and economic options

The self-sufficiency of the milk market in the EU-15 approached 120% in all the years from 1997 to 2002. The market pressure will continue as long as milk quotas exceed demand and the current reform of the CAP will put additional pressure on producer prices, and both effect profitability of milk production. As it is known the intervention prices for skimmed milk powder decline by 15% and for butter by 25% the next three years. At the same time increasing quotas will prevent prices from recovering to the old level. On average the EU-Commissions prospective declines to 24.2 € per 100 kg milk.

Price levels in the candidate countries approached the EU-15 average milk price in the last decade and the price gap declined. But milk prices always have been varying between countries even in the EU-15, where for example in Italy the milk prices have been more than 30% higher than in the United Kingdom in the four years 1999 to 2002 (Figure 5). Regional price differences are expected to continue and influence profitability.

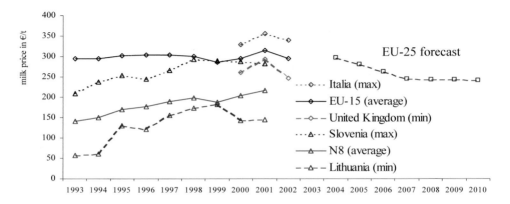

Source EU-Commission and ZMP

Figure 5. Development of milk prices in average for EU-15 and eight new countries and forecast.

The increasing economic pressure will lead to also declining milk quota prices. In Germany for example, the quota prices achieved at the quota bourse have reached a level of 0.4 € per kg, which is only half of the value from two years ago. The new member states can certainly take influence on the structural change by controlling the development of quota prices. If it is the objective to build up a competitive dairy industry, it is recommended to collect all unfilled quotas in the national reserve and to distribute it to expanding dairy farmers. If the demand exceeds the available quotas, of course someone has to decide according to the rules of a common agreement and in biggest possible social consensus. This will help to fulfil the quotas even in a strongly changing environment and in view of the high investments and costs for growing farms.

It can be expected that the decreasing cost for milk quotas and the newly introduced milk premium in the EU-15 can not compensate the reduction of milk prices. On the long run a further rationalisation and structural change of dairy farms can be expected especially in the small structured regions in the old as well as in the new member states. Calculations have shown that new investments in milk production presume minimum herd sizes of 80 to 100 cows even in Bavaria (AGRA-EUROPE 7/2004). Even for farms in Eastern-Germany, where the average herd size already amounts 170 cows (Mecklenburg-Vorpommern) a yield of 9,000 kg is necessary to reach profit after decoupled premiums (result of own calculations).

To become competitive with the EU-15, the necessary structural change in countries with average cow herds below 3 cows will be enormous. These countries have still advantages in some cost factors, for example lower labour costs and lower quota costs and farmers can get state subsidies for investments. The milk prices are expected to approach the community level, especially when the demand for milk products is increasing, as the forecast shows in Figure 6. In Germany already 40% of the milk is used for cheese production. If the average EU-25 consumption of cheese increases from 17.0 kg per capita in 2002 to 18.6 kg in 2010, then milk producer prospective's can not be too bad.

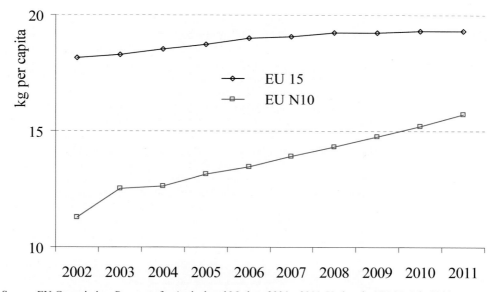

Source: EU-Commission: Prospects for Agricultural Markets 2004 – 2011, Update for EU-25. July 2004.

Figure 6. Cheese market projections in the EU-25, 2002 – 2011, per capita consumption (kg).

Farm income potentials and necessary farm structure adjustments

The subsequent analysis presented uses a method to assess the development of typical farms and shows one example for income potentials and necessary structure adjustments (20 ha farm in Lithuania). The analysis is based on the **E**conomic **A**ccounts for **A**griculture (EAA), on **F**arm **A**ccountancy **D**ata **N**etwork (FADN) data and on the **S**tandard **G**ross **M**argins (SGM) of selected Candidate Countries, as well as additional data and assumptions from other sources (Eurostat, DG AGRI and national sources). Starting from the empirical database, a normative multi-period farm level approach is taken to calculate income variables as well as the equity situation due to the changing economic situation during the accession phase and the still ongoing transition process. The phasing-in-period for direct payments until 2013 is considered. The planning period for simulation and analysis is longer than one decade in order to capture the relevant development. Assumptions have been made for prices after accession, yield increases (1% p.a.) and cost increases (1.5% p.a.).

Assumptions about characteristic attitudes of subsistence and semi-subsistence farms and their households

The subsistence farms are normally relatively small. These farms serve the basic – primarily housing and food - needs for the family and sell additional surplus on mostly local markets. The revenue of this monetary surpluses can be used for family expenditures or farm investment. These payments and investments vary according to natural or economic conditions. In favourable times, more money can be spent for family-living purposes and for investment. The ranking for spending is assumed to be variable. First priority goes to variable factor inputs to assure food production, second to family expenditures, followed by investment in fixed assets such as buildings and machinery. Due to the lack of surety, the banks will give no loans, and investments are the lowest ranked priority. It would not be unusual that in tight financial situations investments are postponed to later periods and the value of fixed assets will decrease over time. If there is not enough liquidity to maintain living standards, then it is assumed that the part of family living expenditures which comes from agricultural activities will be reduced. Because the farms are relatively small, they are often run part-time. In this case other off-farm income or payments from the social security system deliver complementary payments. For subsistence farms the investment gap and income gap is calculated to show economic pressure or the need for external funding.

In conclusion, it can be stated that even in the case of reduced profitability, farming business can continue over a long time and a gradual transformation into part-time farming can be observed. If economic conditions improve or in the case of accession the subsidies increase, then a turnaround could occur and investments and farm size growth could be observed, at least for farmers willing to take their chance in the farming business, while others take advantage of opportunities for (early–) retirement or in other economic sectors and leave the farming community.

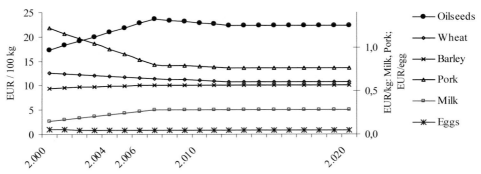

Source: Eurostat; EU-Commission (2002) Analysis of the Impact on Agricultural Markets and Income of EU Enlargement on the CEECs; own calculations

Figure 7. Product price development according to the "CAP-price scenario" in Lithuania.

The scenario **"CAP-prices"** is used to show the isolated effect of a common market in the enlarged EU. Product prices in the CEECs are assumed to adjust to EU price levels[4]. For oilseeds and barley an increase in prices is expected (Figure 7). For Lithuania the increase in milk prices would be most noticeable, while the prices for wheat and pork could decrease.

In a scenario called **"Further structural change with net-investment and farm size growth"** typical farms are treated separately in developing an investment scheme. In the case of lack of own financial resources, the option of net-investment is restricted to the access to bank loans on financial markets or at least depends on state support, for example the gathering of a flat rate of 1,000 € after the approval of a business plan in the cases of semi-subsistence farms.

The growth in area require additional investments at least in machinery and often also in farm buildings. It is assumed that the growth in the net-investment sum could be limited to 1,000 €/ha, if farmers co-operate, and make use of scale effects, and if possible buy used machinery at reasonable prices. In the case of expanding livestock production a minimum of 1,000 € per additional livestock unit (500 € for the animals and 500 € for investment in buildings) is calculated as the net-investment. As a first example, a possible growth path for the case of a typical 20 ha Lithuanian semi-subsistence farm will be described next.

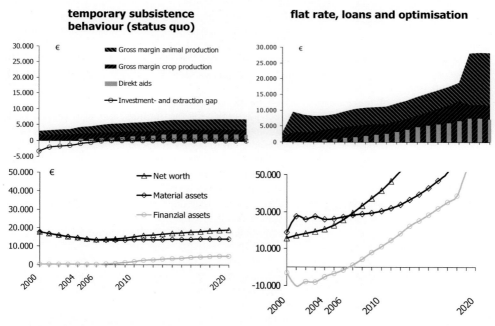

Source: FADN data and own calculations

Figure 8. Economic development of a semi-subsistence farm with 20 ha and 3 cows in Lithuania, depending on different strategies (static behaviour or optimisation approach).

---

[4] Source: Analysis of the Impact on Agricultural Markets and Income of EU Enlargement on the CEECs. March 2002. Prices of potatoes and eggs in the CEECs are assumed to approach price levels according to Member States with low price levels (Source: Eurostat).

Keeping the farm production on a status quo level, the income would not be sufficient to serve the family even with increasing CAP subsidies (Figure 8, left side). The farm can only avoid bankruptcy in the long run, if up to family could reduce financial extractions by 10,600 € in seven years or achieve off-farm income.

A more progressive strategy with a growing area and an extended cow herd is much more profitable. The effects of an additional rental area of 10 ha, which gives a farm size of 30 ha in 2010, and an increase of the cow herd from 3 cows to 7 cows are shown in Figure 8 (right side). Crop rotation is changed to the more profitable cereals, oilseeds and fodder production, increased as necessary to feed the expanded cow herd.

The amount of net worth capital would grow and there would be a stock of accumulated own financial resources to support further operational growth and to consolidate the economic prosperity. There is no doubt that the numerous investment alternatives and further prospects for development, including efficiency improvement by increasing natural yields, could be used for further improvements.

Summary

The milk in the EU-25 is produced by 2.3 Million farmers which are mostly located in the new member states. The structure of milk production is of dual form, some countries produce on large scale (CZ, SK), while others with an average herd size of 3 or less cows face a huge structural change.

Foreseeable developments are a reduction of the small production units and a decrease in own consumption and direct sales. Not only the higher quality standards but also necessary investments will force the structural change.

The integration of the milk production in the new member states into the CAP and its actual reforms will bring increasing direct payments for the farmers, but also reduce milk producer prices. The introduction of the quota system in the new member states is a task to be managed and to the benefit of the farmers who want to invest in the dairy business.

The subsequent analysis presented is based on the Economic Accounts for Agriculture (EAA), on Farm Accountancy Data Network (FADN) data and on the Standard Gross Margins (SGM) of selected Candidate Countries (Lithuania and Hungary), as well as additional data and assumptions from other sources. Starting from the empirical database, a normative multi-period farm level approach is taken to calculate income variables as well as the equity situation due to the changing economic situation during the accession phase and the still ongoing transition process.

The analysis is done for different farm types and size classes in an attempt to cover the development for typical farms (subsistence farms, semi-subsistence farms, extensive dairy/beef farms, medium size cereal farms and large farms). The results generated show on the one hand the economic pressure which is a real burden on farms due to the lack of economies of scale and to low efficiency, but also possible strategies to overcome the current difficulties.

References

Agra-Europe, 2004. Milchwirtschaft braucht dringend Strukturwandel, 16.02.04, Länderberichte, 7: 31-32

EU-Commission, 2002. Agricultural Situation in the Candidates Countries - Country Reports: Bulgaria / Cyprus / Czech Republic / Estonia / Hungary / Latvia / Lithuania / Malta / Poland / Romania / Slovak Republic / Slovenia.

EU-Commission, 2002. Analysis of the Impact on Agricultural Markets and Income of EU Enlargement on the CEECs.

EU-Commission, 2002. Medium-term prospects for agricultural markets.

EU-Commission. Farm Accountancy Data Network (FADN).

EU-Commission, 2003. Accession proposal 27_10_2003 en.pdf

EU-Commission, 2004. Prospects for Agricultural Markets 2004 – 2011, Update for EU-25. July 2004.

EU-Commission: The 2003 Agricultural Year.
http://europa.eu.int/comm/agriculture/agrista/2003/table_en/index.htm

EUROSTAT. Economic Accounts for Agriculture (EAA).

Lithuanian Statistical Office, 2002. Notes from March and May 2002.

Wohlfarth, M., 2004. EU-Erweiterung – Milcherzeugung und –verarbeitung in den Beitrittsländern. ZMP-Nachrichten, Milch – Nr. 8/2004: 7-10

ZMP, 2004. Agrarmärkte in Zahlen - EU 2003 sowie EU und EU-Beitrittsländer, Bonn.

Appendix

The milk market in the enlarged EU

	Year	unit	EU-25	EU-15	in %	N10	in %	Source
Area		000 km²	3.891	3.154	81%	737	19%	[1] S. 169
Population	2002	Mio.	454,4	379,6	84%	74,8	16%	[1] S. 169
Milk producer	2000	000	2.323,0	734,0	32%	1.589,0	68%	[6] S. 2
Cow stock	Dec. 2002	000 head	24.860	19.940	80%	4.920	20%	[2] S. 171
Average cow herd size	2002	cows/farm	15	29	193%	3	20%	[6] S. 2
Milk yield per cow	2001	kg/year	5.730	6.050	106%	4.050	71%	[2] S. 170
Milk production	2002	Mio. t	143,6	121,8	85%	21,8	15%	[2] S. 171
- Forecast EU-Commission	2010	Mio. t	145,2	122,7	85%	22,5	15%	[3] S. 44
Milk production	2001	kg per capita	321	326	102%	300	93%	[2] S. 173
Milk quota	2005	kg per capita	306	316	103%	253	83%	[2] S. 173
Fresh milk consumption	1999	kg per capita		105,1				[4] S. 467
Cheese consumption	2002	kg per capita	*17,8*	18,8	106%	*10,7**	60%	[2] S. 180
- Forecast EU-Commission	2010	kg per capita	*18,6*	*19,3*	104%	*15,2*	82%	[7]
Butter consumption	2002	kg per capita	*4,3 **	4,4	102%	*3,6**	84%	[2] S. 177
- Forecast EU-Commission	2010	kg per capita	*3,86*	*4,03*	104%	*2,97*	78%	[7]
Milk price	2002	€/100 kg		29,50		17,90 to 29,63		[2] S. 173
- Forecast EU-Commission	2010	€/100 kg	*24,1*					[3] S. 24
- Forecast Prof. Weindlmaier	2007/08	€/100 kg	22,00					[5] S. 32

Sources:
[1] ZMP: Agrarmärkte in Zahlen, EU 2003;
[2] ZMP-Marktbilanz Milch 2003;
[3] EU-Commission: prospects for Agricultural Markets in the EU 2003-2010;
[4] Stat. Jahrbuch über E_Lu.F. 2001;
[5] AGRA-EUROPE 7/04, 16.02.04, Länderberichte;
[6] AGRA-EUROPE 10/04, 8.03.04, Markt+Meinung;
[7] EU-Commission: Prospects for Agricultural Markets 2004 – 2011, Update for EU-25. July 2004.

Farming with quota

Bram Prins*

Partner in family farm, Head of IFCN Cream Club, Freelance advisor in strategic management, The Netherlands

Introduction

Bram Prins will tell about the goals of his home farm, experiences with the Dutch quota system as introduced in 1984, the consequences of the quota system for the EU market and for the development of the home farm and he will present some thoughts about the future.

Mission of our farm

The mission of our farm is to produce food in a for farmer and society desired way. The farmer needs to find a balance between his own ambitions, attitude and qualities and the wishes and regulations coming from the society as a whole. As farmers community we have to look at both the market and the society. We need to learn to look outside our own borders, the borders of our farm and home village. It is not enough to look simply at the price of a product, because the price is a result of the developments in the market and in the society together.

Figure 1. Holstein-Friesian cows on the grassland.

*This paper is written by Dr. Abele Kuipers on the basis of the presentation of Mr. Bram Prins at the workshop in Bled, Slovenia and on an interview held later.

The goals of Bram Prins can be illustrated as follows:
- External targets:
  - From product oriented to market oriented and society oriented
    Triple P (Profit – People – Planet)
- Internal targets: ***Triple L***

Labour productivity

Labour time   Labour pleasure

As internal target Bram Prins focuses on the Triple L: Labour productivity, required Labour time (versus spare time) and Labour pleasure. To continue farming in an economical sound way, labour productivity has to increase. This asks for a continuous effort in applying new knowledge and new techniques. Labour time needed to perform all tasks should preferably not increase. When more and more time is needed per person to fulfil all duties on the farm, the life style of this farmer does not fit in the society surrounding him anymore.

The farmer and his family may become isolated. Another internal criteria to be considered is labour pleasure. Without some pleasure in doing the daily work, the farmer becomes a less friendly person to his neighbourhood. This is especially troublesome when he has to deal with more employers on the farm. A bad humour is deteriorating the cooperative attitude of people.

Some experiences with Dutch Quota System

The quota system as implemented in The Netherlands in 1984 can be characterised by:
- same system all over the country;
- free market: no govermental regulation;
- reference quota amounts are linked to the ground;
- different quota traders: with non-official agreements.

There is a lot of discussion in our country about the new phenomenon of quota traders. The feeling is that these marketers control more or less the quota transfer market in such a way that quota prices are manipulated. This has resulted in the strong wish of the Farmers' Organisation (LTO) to make the quota market more transparent and regulated. This may hopefully result in lower quota prices than currently exist.

We can observe different phases in the attitudes towards the quota system:
- Before the quota system was actually introduced: strange reactions of farmers. Some farmers invested heavily as reaction to a restricted quota regime to come. Other farmers tried to neglect or avoid the system.
- Quota system just installed: no growth of the farms. Farmers were thinking that all developments were blocked with this system of quota.

- After some years were passed: farmers started to pick up again the strategy to enlarge their farm. However, rising quota prices diminished the marginal benefits of enlargement of the farm business.
- After a longer period: higher cost-prices emerged and the advantages of the system are lost.

Consequences of quota system for EU market and our farm

The effects of the quota system on the EU market is considered by Bram Prins as minimal in these days. If the quota system would disappear, the market would only react slowly. Prins argues that the market does react much more strongly at political changes than at reactions of the farmers' community. Also currency differences between countries in "old" EU did influence the milk price significantly. The price is sensitive towards these currency gaps. This also applies to the CEE countries. They have other currency than the EU. Prins believes that with a quota system we need barriers on the EU borders. Restrictions within the Union should be protected by restrictions in imports.

How is family Prins dealing with the quota system?

When starting with quota:
- consequences for my father
  - he was member of the regional committee who assigned quota to farmers
  - he experienced a period of no development of the farm
- consequences for the farm:
  in the period from 1984 – 1990:
  - use of jersey crossings at our farm to increase fat percentage
  - breeding of heifers for export purposes
  in the period 1990 – 2003:
  - we increased our quota amount by 60%
  in 2003-2004:
  - we were running as farm in problems,
  - the quota price is too high,
  - we shift our attention now to other issues, like cost level, etc.
  Our conclusion: too much emphasis has been on requiring more milk quota.

Indeed, the quota system did influence our management

We utilized more the services of the extension service:
- First we looked at simple improvements and at how to fill the barns again. We moved from the strategy of enlargement of the farm to the strategy of consolidation. Also awareness for the level of expenses was growing.
- Later on we increased the quota amount with hardly any calculations as base.
- But in present time: Business planning is done and quota appears too expensive to invest in!!

Future of the quota system and our farm

Bram Prins describes this by a few statements and by raising some questions:
- For me as farmer the quota system has no value
- Important is under which circumstances to go out of the system of quota

- What will be the influence of a situation without quota on the cost-price?
- What will be the expected growth of the milk volume when the quota system would disappear?

These issues and questions raised will determine the future of the quota system. Prins believes that "we are running out of the system gradually". Because of lower milk prices in Western Europe, some countries like France, Ireland and Germany have already reached a situation in which they are not utilizing their national quota amount fully. Also in the Central and Eastern European "new" EU countries, there exists a situation of non-utilization of national quota. Considering this, Prins believes that the step to a free market system is becoming more and more a realistic option.

The policy of the Prins family to deal with the future is:

- Be consequent: stick to the chosen strategy.
- No emotions in decision making, but before decisions are really made proper calculation work needs to be done first.
- Try to realise a growth that is related to the growth in other sectors and search for income from diversification, like in our case from biogas.

My conclusion as Dutch farmer:

Quota system is running on its end and we will be happy when this happens!!!

Canada's forty years of milk supply management manage quotas for farm profitability

Robert L. Lang

Former director International Livestock Management Schools- Canada, Canadian Livestock Genetics Association

In 1965 Canada had 123,000dairy farms with 2,500,000 cows that produced some seven million tons of milk. The production pattern was highly seasonal, especially in the provinces of Quebec and Ontario that had more than 80% of the cows. Milk quality was varied with very segregated prices of milk for industrial versus fluid markets. Producers were dominated by processors, especially the few farms that produced milk for the fluid market. Federal and provincial subsidies accounted for any profit derived from the production of milk and producers were beholding to the political parties to continue the "consumer subsidy" programs. The dairy industry was dominated by a "cheap food policy" that deprived dairy farers from earning a competitive return on their intellect, investment and hard work.

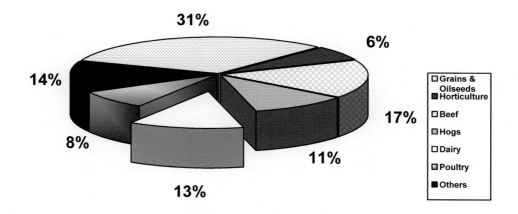

Figure 1. The Canadian Dairy Industry is the third largest sector of the Canadian Agri-Food economy, after grains and red meat.

In 1965 Canadian food policy changed. Enabling legislation was introduced that allowed each province to manage a share of the national industrial milk supply. Provincial "quotas" were allocated. The provinces were allowed to institute raw milk sale monopolies that were restricted to producers licensed by the Province to produce milk for wholesale purposes. This was still a segregated supply system with cream, industrial milk and fluid milk priced differently due to widely differing qualities and with seasonal milk prices a normal practice.

Gradually, Provincial governments delegated full monopolistic control of milk production and wholesale supply to producer associations. These "Milk Marketing Boards" set policy and managed the elements of the raw milk production system within a province.

Quotas initially had no financial value. Quickly, however, the higher priced fluid quota generated value and could be traded amongst producers within a province. Fluid prices were a product of the consumer forces. Industrial milk was priced according to a "Formula Pricing Methodology" that was designed to allow the top 50% of the producers to derive a profit, income over a break even price. In reality, as the system evolved every producer improved milk production efficiency, milk quality, reduced seasonality of production and moved to bulk storage of cooled milk. Specialty dairy farms became the norm. Prices escalated and government subsides gave way to market forces.

Today Canadians produce only one quality of milk. All producers are paid by the utilization of milk in the market. They receive their payment routinely within 20 days of the end of the month. Milk is paid at the top quality rate with deductions for less than top quality and safety. Quotas are met on a monthly basis instead of annual. There are no subsidies and the domestic market is the sole market or Canadian milk. Dairy farms are now the most consistent and stable net farm income earners in the Canadian agricultural system.

Table 1. Trends on the farms in Canada.

	1970/71	2002/03	Difference , %
Number of farms	122,194	18,050	- 85%
Number of milk cows (thousand heads)	2,389	1,075	- 55%
Cows per farm	19.5	59.6	+ 206%
Volume of milk produced (million hectolitres)	77.05	74.29	- 3.6%

Changes have occurred however. There are now less than 19,000 dairy farmers milking 1.1 million cows producing almost 8 million tonnes of milk worth $.2 billion dollars at the farm gate (Table 1). Average herd size is now over seventy milking cows, almost one hundred hectares and on average 430,000 litres of milk is produced annually per person working on the farm. In order to produce and market milk the producer must own "quota". This costs about $28,000 per kg of fat marketed daily. At 9,717 kg milk at 3.7% fat each cow requires 1.2 kg fat per day and the producer invests more than $33,000 per cow milked for the right to produce and market milk.

The lesson learned in Canada is that "Financial survival means quota management which means producing enough fat to use every gram of quota every day of the year". Although fat is approximately 40% of the value of a litre of milk and protein is another 55% of the value, management of fat yield and use of every gram of quota is necessary to service the cash flow and optimize the farm income.

The next lesson learned is that producers must constantly invest time and energy in the life-long pursuit of applied knowledge to ensure cows reproduce regularly, produce to their maximum ability without succumbing to stress and that feeds grown and harvested are a fit with the cattle's ability to use the feed and efficiently convert these to nutrients to support the growth, maintenance, production and reproduction requirements of the various stages of life the animal goes through.

The major lesson learned is that regular production of milk and components to the fullest use, but not over use of the quota available is necessary to survival in the modern world of supply management.

Through the International Livestock Management Schools, the Canadian dairy cattle genetics cooperatives provided transfer of the supply management technology to our trading partners. Since the late 1980's, countries including Turkey, Russia, Ukraine, Belarus, Estonia, Latvia, Lithuania, Poland. Romania, Slovenia, Serbia and Bulgaria have participated in exchanges that included the study of supply management. Extensive training and technical discussions in these countries with visiting Canadian extension workers and producers have led to a number of observations regarding the pending inclusion of many of these countries into the European Union milk supply management system.

The CEE dairy producers need to avoid what seems to be an emerging adversarial attitude to dairy quotas. It appears that the total focus is on avoidance of over-quota production rather than the balanced approach of meeting 98 to 105 of the allotted quota. Milk income at the CEE farm level seems overly fascinated with the potentials of EU government subsidies, the eventually fatal and addictive "consumer subsidy" approach to keep cheap food on the market rather than fostering strong, competitive milk production through producer driven marketing. The prevailing attitude of producers that the government is the market for milk is eventually non-sustainable and the loser is the producer. Every milk producer in the CEE needs to focus on the true market for their output: consumers in the CEE and in the European Union. Production of safe, long shelf life affordable milk will sustain the future, not short term political interventions. Producers cannot abandon the marketing of their livelihood products to some other agency such as government or the processing industry.

There is a need to change the mind set to a "win-win" situation for producers, processors and consumers. The pending system seems to be heading into a "winner-loser" situation, with producers likely to be the financial losers.

Low input dairy production seems to be fashionable in the CEE. This system is based on seasonal production (milk on grass), low performing cows so that stress will not adversely affect reproduction, manufacturing milk production supported by public subsidies and export of industrial milk products sales.

Production of economically produced, safe milk in synchrony with the processing demands of the industrial sector requires a broad range of knowledge of many technical and management disciplines. Public and private technical extension workers supporting these farmers needs require management competency. This includes practical skills application and the ability to successfully transfer these skills to the dairy farm situation. Unfortunately, it seems that the EU entry situation has focused on training extension workers as desk-bound paper processors instead of production management oriented change specialists. While farm training is constant and the need is always high, the focus of these early days of quota management in the CEE requires that great emphasis be placed on training well qualified extension workers in the applied technical and people dynamics needed to lead traditional producers into a new, more food safety aware and demanding market.

The best advice from our long time experience with milk quota management is for the producer, processor and consumer to cooperate and find ways that all three sectors can win in the production, processing and utilization of milk products. The availability of modern, high performance dairy genetics that are in concert with similar performance plant genetics provides the knowledgeable and competent dairy producer the opportunity to produce competitive financial returns to any other commercial endeavour. Producing effective levels of milk with high quality and annual reproduction requires a great deal of skill applied diligently every day by the farmer. Disease and stress prevention rather than a focus on cure is the knowledge key that leads to dairy profitability.

The discipline to work together with these wonderful cattle and provide highly nutritious food that is the source of financial sustainability for the CEE dairy industry has to come from within each producer. Extension workers who can foster the continued development of the efficiency of milk production by adoption of ever improving technology by serious dairy farmers are the key to ensuring the future of dairy production in the CEE, no matter what the future politics of global dairy production throws at the industry.

Figure 2. Typical Canadian farm.

Country reports

Milk quota and farm management in Republic of Estonia

Katrin Reili and Kalev Karisalu

Ministry of Agriculture, Veterinary and Food Board, Vaike-Paala 3, 11415 Tallinn, Estonia

1. Introduction

The first year of milk production quotas effective in Estonia lasted from 1 April 2003 till 31 March 2004, constituting the first stage of the application of the milk quotas mandatory in the European Union (EU). In this first year the system was applied in a less complicated form than that in the EU, which means that during this year no fines were charged for exceeding the quota, and furthermore the agricultural producers did not have a possibility of buying and selling quotas. The objective of the application of the milk production quota at the first stage was familiarisation of the producers and processors with the quota system, and accumulating experience for quota administration. The second stage started with the accession to the EU at 1st April 2004, and from that time on the EU system is applied to the full extent.

1.1 Distribution of quotas for the first quota year of 01 April 2003 to 31 March 2004

For the first quota year the Government established the national quota at 900,000 tons, 90,000 tons of which is the direct marketing quota, and 810,000 tons is the delivery reference quota.

The dairy farmers whose dairy cows were registered in the ARIB's animal register as at 15 December 2002 had right to apply for the milk quotas. In order to receive the quota the farmer was to be an undertaking, or was obligated to register himself or herself in the tax office, or the commercial register as a sole proprietor (FIE) or company before 15 January 2003.

ARIB calculated the volume of the quota for producers by multiplying the number of the dairy cows of each producer, entered in the register as at 15 December 2002, with the average milk production during the reference period of the dairy cows under the performance monitoring. The reference period was the period between 01.04.2001 to 31.03.2002. During the reference period the average milk production in Estonia per cow under the performance monitoring was 5,591 kg, and the average fat percentage was 4.31%.

Starting from the implementation of the milk quota system a producer, who has not been granted a quota, does not have the right to sell the milk he has produced. However, producers who do not sell milk but use it in their own household instead, are not required to apply for quota.

ARIB send the pre-completed milk quota application forms to 7,120 dairy farmers, and offered them the milk quota equivalent of 622 thousand tons for their 111,315 cows (Table 1). 2764 dairy farmers returned the completed application and requested a milk quota.

733 dairy farmers applied for an increase of their milk quota as stated on the pre-completed application form by reason of having in-calf heifers or above average milk production rates. After on-site verification of the information and verification of the data at the Animal Recording Centre, an additional quota of 114 thousand tons was allocated to those 725 producers (their aggregate quota reaching 412 thousand tons). 1432 producers, amounting to the aggregate quota of 200 thousand tons, consented to the offered quota, whereas 509 producers requested that the quota be reduced.

Quotas were not allocated to 4356 dairy farmers, whose aggregate herd amounted to 10,083 cows. Obviously, the majority of those who refrained from applying for the quota were small dairy farmers who use the produced milk in their own households.

Table 1. Application for milk quotas per county*.

County	Dairy farmers to whom the quota application was sent	Dairy farmers who applied for the quota		Number of animals for whom quotas were granted	Volume of the granted quota (t)	incl. direct marketing quota (t)	incl. delivery reference quota (t)
Harju	326	154	47.2%	7 169	39 898	6 101	33 795
Hiiu	143	73	51.0%	723	4 008	950	3 058
Ida-Viru	171	74	43.3%	2 254	12 267	1 827	10 440
Järva	319	187	58.6%	18 559	116 529	1 591	115 028
Jõgeva	597	209	35.0%	11 542	68 470	1 786	66 683
Lääne-Viru	413	191	46.2%	12 932	75 898	722	75 177
Lääne	386	129	33.4%	3 276	16 048	1 276	14 772
Pärnu	879	374	42.5%	11 904	68 096	3 621	64 474
Põlva	301	115	38.2%	6 805	31 122	1 094	41 028
Rapla	508	253	49.8%	8 059	47 526	1 723	45 083
Saare	676	302	44.7%	5 755	30 228	993	29 235
Tartu	697	162	23.2%	6 696	39 526	1 669	37 857
Valga	619	138	22.3%	3 643	19 148	1 212	17 936
Viljandi	526	224	42.6%	7 989	44 951	2 636	42 315
Võru	559	179	32.0%	3 658	20 962	1 106	19 856
Total	7120	2764	40.7%	110 964	645 675	28 217	617 457

*Data from Estonian Agricultural Registers and Information Board (ARIB)

The first quota assignment took place on 14 February 2003, i.e. before the beginning of the first quota year, when milk production quota was allocated to 2666 producers for 110,964 dairy cows, and the aggregate milk production quota reached 645,675 tons, being higher than the initial quota offered to the producers by 23 thousand tons. 98 dairy farmers of those who submitted the applications, were not granted quotas mainly for the reason that they had failed to register themselves as undertakings by 15 January 2003. 95.6% of the production quota allocated, i.e. 617 thousand tons, constituted the delivery reference quota, and the remaining 28 thousand tons – the direct marketing quota.

By counties the milk quotas were most actively applied for in Järva county where 58.6% of all the applications sent out were returned (see Table 1). It was followed by Hiiu county (51%) and Rapla county (49.8%). The largest quota was allocated to Järva county, followed by Lääne-Viru and Jõgeva counties. The highest direct marketing quota was allocated to Harju county, and the highest delivery reference quota was allocated to the dairy farmers of the Järva county.

1.2 Quota performance

By the end of the first quota year 491,740 tons, i.e. 76.6% of the entire milk quota was produced (see Table 2), 12,148 tons of which (i.e. 54.6% of the allocated direct marketing quota) constituted the performance of the direct marketing quota, and 479,601 tons (i.e.

77.4% of the allocated delivery reference quota) constituted the performance of the delivery reference quota. 28.7% of the entire national milk quota remained in the reserve.

Table 2. Milk production quotas as at 31.03.2004.

Milk production quota (t)	900,000	Direct marketing quota (t)	90,000	Delivery reference quota	810,000
Allocated quota	641,645	Allocated direct marketing quota	22,249	Allocated delivery reference quota	619,396
Performed milk production quota	491,749	Performed direct marketing quota	12,148	Performed delivery reference quota	479,601
National reserve	258,355	Reserve	67,751	Reserve	190,604

The change of the allocated quota is caused by the change of the number of quota holders during the year, i.e. the new applicants and those who surrendered their quotas during the year. By the end of the quota year there were 2428 quota holders left (see Table 3). The main reason for surrendering the quota was dissolution of the operations. However milk production quota was allocated to 93 new producers, so the total number of opters-out was 331.

Table 3. Dairy farmers who held quotas as at 31.03.2004.

County	Dairy farmers who held production quotas as at 31.03.2004			New applicants during the quota year
	Total	incl. FIE	inc. companies	
Harju	138	108	30	2
Hiiu	45	42	3	2
Ida-Viru	64	56	8	1
Jõgeva	174	151	23	5
Järva	176	147	29	10
Lääne	103	96	7	3
Lääne-Viru	168	115	53	4
Põlva	126	109	17	22
Pärnu	322	294	28	8
Rapla	229	208	21	7
Saare	273	255	18	9
Tartu	143	121	22	8
Valga	114	107	7	3
Viljandi	191	168	23	3
Võru	150	140	10	4
Undefined	12	10	2	
Total	2428	2127	301	93

The largest number of quota holders is still in Pärnu county, which was followed by Saare and Rapla counties. The largest number of new quotas was applied for in Põlva county, followed by Järva and Saare counties. New quota holders appeared in all counties during the year. Põlva county was the only one where the number of new quota holders exceeded that of the opters-out (see Figure 1), in other counties the number of opters-out was larger, whereas the number of them was the highest in the Pärnu county, and the highest percentage was reached in Hiiu county where within the year the number of quota holders decreased by 38%.

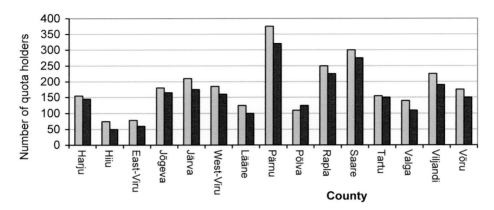

☐ Dairy farmers, who applied for the quota before the beginning of the quota year.
■ Dairy farmers who held the production quota as at 31.03.2004

Figure 1. Changes in the number of quota holders from 01.04.2003 to 31.03.2004.

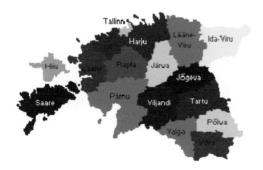

Figure 2. Map of Estonia with 15 counties.

During the quota year the number of cows covered with milk quotas had been reduced to 98 899 (see Table 4). The most significant reduction took place on the account of opting-out of small dairy farmers with less than 10 dairy cows. Among larger herds the number of opters-out was small or virtually non-existent.

Table 4. The number of dairy cows and the average fat percentage at the end of the quota year.

Size of the herd	Number of dairy cows as at 31.03.2004	Actual average fat percentage
1 ... 10 cows	6,702	4.31
11 ... 50 cows	14,824	4.22
51 ... 100 cows	6,757	4.15
101 ... 500 cows	49582	4.12
Above 500 cows	21,034	4 12
Total	98,899	4.24

While during the first year the allocation of the milk quota was based, among other factors, on the average fat percentage of the milk produced by the dairy cows under the performance monitoring, which was 4.31%, the actual individual fat percentage will be taken into account in determining the quota for the subsequent quota years. The last column of Table 4 represents the actual average fat percentage by different herd sizes.

In spite of the fact that during the year the largest number of opters-out represented dairy farmers with smaller herds, the majority of quota holders are still producers with herd sizes between 1 to 10 cows (see Figure 2).

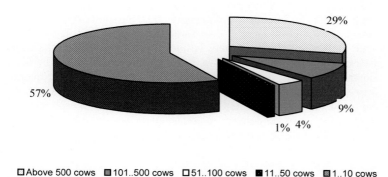

Figure 2. Distribution of quota holders by the size of their herd during the quota year of 2003/2004.

While the smaller herds still make up the majority of recipients of milk quotas, they have nevertheless been granted nearly the smallest share of the quota itself (see Figure 3). Over two thirds of the quota is held by owners of large hers.

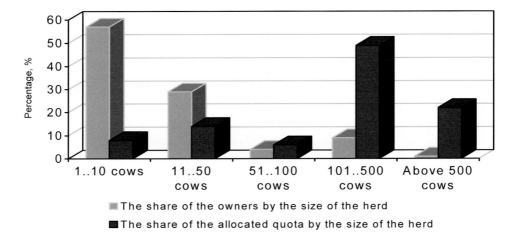

Figure 3. Distribution of quotas among the dairy farmers as by the size of their herd.

Generally the allocated milk quotas have not been exceeded. The only exception is Saare county, where during the quota year the direct marketing quota has been exceeded by 70%. While the average quota performance reaches approximately 77%, the above-average quota performance was demonstrated by Jõgeva, Järva, Põlva, Tartu and Lääne-Viru counties. With respect to the performance of direct marketing quotas, besides Saare county where the quota was exceeded, the highest results were shown by Järva, Põlva and Tartu counties. The dairy farmers of Jõgeva counties produced only a quarter of the direct marketing quotas, demonstrating the lowest production performance. In general at least two thirds of the allocated delivery reference quotas were performed in the counties, the only exception being Hiiu county (61%). The largest share of the delivery reference quota was performed in Põlva, Järva and Tartu counties (in all these the performance exceeded 80%).

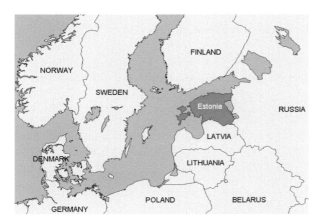

Figure 4. Map of Estonia.

2. Structural changes due to EU-quota system and premiums

a. How is the structure of the dairy sector expected to change from now (2004) till over 5 years (2010)?

The number of cows is 118.2 thousand in 2004 and it is expected to reach to 120 thousand in 2010. The number of dairy producers will decrease quite remarkable in coming years, but all this will happen on account of small producers. It can be foreseen that there will dominate 400 cow herds will dominate in the future (all stables that are build or reconstructed have this size). The average production per recorded cow is expected to rise to 6000 kg in 2005 and to 7000 kg in 2010.

b. What effect will the quota system have on the other animal sectors?

The quota system likely shifts farmers to choose if they want to continue to produce and market milk or not any more. Producing of milk for self consumption will disappear and many former dairy farmers will choose some alternative production type, like fattening of bulls or keeping of beef breed cattle or sheep. The mixed farming is very wide spread in Estonia and we expect it to endure.

c. Do you expect regional differences in development of agriculture? Do you have regional quotas or one national milk quota?

It can be observed that some differences in development of agriculture in Estonia will occur, as the soil conditions are more suitable for agricultural production in the central part of the country. There are also situated the biggest crop producers and cow herds. Nevertheless, there is only one national quota in use in Estonia.

d. Do you expect intensification or extensification of dairy farming with quota?

It is expected that the trend of intensification of dairy farming in the coming two-three years will continue (because we have some room between the national quota and the actual amount of marketable milk at the moment, and probably we are getting a special restructuring reserve as from 2006 onwards), but in the long run the use of the quota system will lead the dairy farming sector in the direction of extensification.

e. How large is the national reserve of quota in your country? Which farmers' groups will be considered eligible to apply for more quota?

There is not set any obligatory amount regarding the national reserve at the moment. Our quota reserve is formed from quantities that are not allocated as individual quota to the producers.

The producer can apply for a bigger quota, if:
- there are free quantities in the quota reserve;
- the producer did fulfil his quota at least 90% in previous quota year;
- the average productivity of cows has been risen, and/or the number of dairy cows has been increased.

f. What was the average milk price from 2000-2004? How do you expect the milk price to develop in your country or region after the entrance of the EU?

Table 5. Average milk price 2000–2004*.

	2000	2001	2002	2003	2004 (6 months)
Average milk price (EEK/100kg)	271.9	319.7	280.0	288.2	382.5
Average milk price (EUR/100kg)	17.4	20.4	17.9	18.4	24.4

Source: *Data from Estonian Agricultural Registers and Information Board (ARIB)

The milk price is expected to stabilize at slightly increased level in coming years.

3. National support

Do you have in your country specific national subsidies to support dairy or beef or suckler cows or sheep farmers? Are those subsidies changing after the accession to EU?

Estonia has paid an annual direct support for keeping of dairy cows since 1998, for sheep farmers the direct aid has been paid since 1999 and for suckler cows since 2001. Annual support for bulls has been paid in 1999 and 2003. Until 2004 there was a performance testing requirement for dairy cows to get premium.

Principally there are no big changes in support requirements for dairy and suckler cows and sheep after accession to the EU. However, after accession the direct support is also paid for calves from the age of 1 month.

4. Effect of milk quota on farm management

a. Do you expect the farmer to change his management under quota: Which management practices will be adapted?

Because of using this milk quota system such a little time, it is difficult to say now in what direction farmers will change the management due to quotas. Probably there will not be any remarkable changes at all (at least for the next few years).

b. Do you expect the farmers and breeding organisations to use other breeds and / or breeding goals under the quota system?

We do not expect any major changes in using breeds and/or setting breeding goals under the quota system. The share of Estonian Holstein breed cows is slightly increasing all the time, currently we have about ¾ of all cows in Estonia representing that breed and this number will reach to 80% in the near future.

c. Do you have in your country (planned or in action):
- Free transfer of quota from farmer to farmer? or
- A national quota exchange bureau? or
- Quota linked to the land / to acreage? or
- A system in which farmers who stop are delivering their quota to a central agency?

In Estonia there is no free transfer of quotas at the moment and it is not planned to introduce it at least in the first couple of years. A transfer of quotas is only possible by way of handing over (by selling, renting or inheriting) the farm or part of it from one producer to the other.

d. If there is some kind of transfer of quota, what level of quota prices do you have at the moment or do you expect to have?

There is no free transfer of quota in our country, so no quota price exists.

e. Is the instrument of leasing introduced in your country?

Quota can not be leased. Only whole or part of farms can be transferred from one person to the other.

f. With the allocation of individual quota to farmers, which reduction percentage was used?

There were specific rules in force for the first quota year (1 April 2003 – 31 March 2004) in Estonia. According to this, if the producer did not fulfil the quota, it was decreased after the quota year (but before the accession date 1 May 2004) as follows:

- If the quota was unfilled 10-20%, the decreasing rate was 50% of the unfilled quota;
- If the quota was unfilled 21-30%, the decreasing rate was 75% of the unfilled quota;
- If the quota was unfilled over 30%, the decreasing rate was 100% of the unfilled quota.

As a result of those decreases the amount of allocated individual quotas reached in case of deliveries to 515,720 tonnes (96% from Estonian national delivery quota) and in case of direct sales to 11,490 tonnes (13% from national direct sale quota).

g. What are the largest cost factors on a dairy farm in your country?

Three major cost items of milk producers – feedstuffs, wages and social tax, and fuel – form on average about 70% of the total production costs in Estonia. The share of feedstuffs is the largest for farmers with 11-50 cows (ca 50% of total costs) and the lowest for farms with herds over 100 cows (slightly over 40%). The share of wages and social tax is much bigger in large farms (up to ¼ of total costs). The share of fuel costs is much smaller for large-scale producers than for smaller holdings.

5. Organisation of extension and extension needs

a. How is extension organised in your country?

Agricultural information (dissemination of state information) is organised through the extension (information) centres of Estonian Chamber of Agriculture and Commerce (ECAC). There are 15 of such information centres in Estonia (one in each county) – mainly housed in the offices of local farmers' unions, where certified agricultural advisors or employees of farmers' unions serve their clients.

The Centre's task is to disseminate information necessary for rural entrepreneurs and the food industry and notify the Ministry of Agriculture about the information needs of rural entrepreneurs and the food industry. The main task of the Centre is to create opportunities for rural entrepreneurs to receive the necessary information in a timely and understandable manner.

This system is backed with an Internet-based agricultural extension system: the PIKK portal. PIKK-portal collects and accumulates information necessary for rural entrepreneurs, including information concerning the environment, rural development measures and processing agricultural products. In the portal one can express one's opinion, ask questions and put up purchase and sales adds. Since the target group of the portal are mainly Estonian rural entrepreneurs and food processors, www.pikk.ee is available only in Estonia.

b. Does extension play a role in explaining the quota and premium system to farmers?

Yes, the explaining of the quota and premium systems is included as one of the activities of ECAC extension centres. These questions (as well as the activities of the Milk Quota Council) have been the object of discussion and enlightenment on the PIKK-portal.

c. Which special extension needs / questions do you expect concerning quota and premiums?

The producers need more often and more precise information about the running quota profile (both, their own and national quota).

d. How do you expect your extension service or other institutions to tackle the new situation?

We feel that the ECAC Information Centre and PIKK-portal have done a good job in informing farmers and other rural entrepreneurs. Therefore we think that these two channels should be used when covering any future changes in the milk quota and premium system.

6. The biggest challenge for the dairy industry in years ahead in Estonia

The biggest challenge for the Estonian dairy industry will be in finding and reaching a reasonably balanced level between the possibilities coming from our comparative advantage in milk production (low production costs, etc) from one side, and the needs coming from the under financing of the sector during last decade from the other side.

Situation of dairy farming and experiences in applying quota system in Hungary

Gyula Meszaros[1] and M. Ignácz[2]

[1]*Livestock Performance Testing Ltd., Dozsa Gy. Str. 58, P.O.Box 258, 2101 Gödöllő, Hungary*
[2]*Agricultural and Rural Development Agency (ARDA), Hungary*

Cattle husbandry in Hungary

Hungary is situated in the centre of Europe (if Europe extends from the rocks of Gibraltar to the Ural mountains - as President DeGaulle stated once). Hungary is home land for ca. 10 million citizens.

Figure 1. Map of Hungary in Europe.

Cattle farming in Hungary – in terms of number of farms involved - is declining, though our country has traditionally excellent geographic and climatic conditions for large and small farms. The statistical structure of our cattle industry is unique:
- less than thousand large farms with about 600 cows/farm and
- more than 20 thousand small farms with about 7 cows/farm.

Table 1. Farms with cattle and cows.

Years	Cattle		Cows & Pr. Heifers		Cattle/Farm	
(December)	Companies	Private farms	Companies	Private farms	Companies	Private farms
2000	927	45,220	847	34,079	586	5.8
2001	804	38,616	743	30,525	618	7.4
2002	797	35,279	740	27,490	619	7.8
2003	860	31,413	783	23,633	568	7.2

Source: KSH

The cattle and the cow population have decreased:
- the number of cattle became less than half,
- the number of cows has become almost half of the number in the years of 80's.
- moreover the cow/cattle ratio has also changed unfavourably (Table 2).

Table 2. Changes in the cattle and cow population during 3 generations.

Years	Cattle (in 1000)	Cow (in 1000)	Cow/Cattle (%)
1986-90	**1,650**	**658**	**39.88**
1991	1,420	559	39.37
1992	1,159	497	42.88
1993	999	450	45.05
1994	910	415	45.60
1995	928	421	45.37
1996	**909**	**414**	**45.54**
1997	871	403	46.27
1998	873	407	46.62
1999	857	399	46.56
2000	805	380	47.20
2001	783	368	47.00
2002	**770**	**362**	**47.01**
2003	714	337	47.20

Source: KSH

The proportion of dairy cows is 88% of the total cow population as shown in Table 3.

Table 3. Distribution of cow population according to their type (December 2003).

Total cow population	337,000
- Dairy cows	298,000
- Beef cows	39,000

Source: KSH

The domestic milk production and milk consumption shows changing figures during the last – more than 30-years (Table 4).

Table 4. Milk production and consumption.

Years	Production	Import	Export	Consumption
	Million litres			litres/capita
1970	1,896	182	182	108
1980	2,478	65	236	204
1990	2,336	25	487	203
1995	1,936	83	238	160
1998	2,061	58	419	163
1999	2,051	138	431	160
2000	2,094	178	413	171
2001	2,095	132	481	161
2002	2,100	165	428	160
2003	1,950	208	435	169

The total milk production showed a consistent improvement between 1970 and 2002 but the hectic way of change in milk consumption was ended with a less than a healthy result. In terms of comparison with our home standards in the previous years as well as with the averages of other countries, the milk consumption in Hungary is rather low.

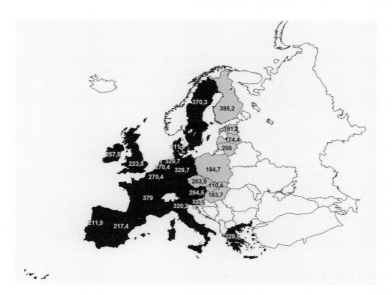

Figure 2. Milk consumption in the "**old**" and in the "**new**" member countries of EU (kg per capita in 2000).

Quota system

Main characteristics of Hungarian milk quota system are:
- Established in 1996.
- Used to be operated by the Milk Product Council. Necessary changes were made from 1 May 2004 on.
- Recently operated by the Agricultural and Rural Development Agency (ARDA) with the assistance of the Livestock Performance Testing Ltd.
- Total quantity of national quota: 2 billion liters (before accession)
- Total quantity of national quota:
  - 1,947,280 tons (after accession):
    - 1,782,650 tons for deliveries and
    - 164,630 tons for direct sales
- Total quantity of allocated quotas to producers: ~ 1,8 billion liters. The allocated individual quotas are properties of the producers
- Individual quotas can be used from 70 – 100%
- No comparison of fat contents – no correction of quota fulfilment before the accession
- Since 1 May 2004 the recording of quotas and accomplishment of individual quotas in kilograms is introduced.
- Project of evaluation and determining of individual representative fat contents executed by the Livestock Performance Testing (LPT) at the end of last year – the basis of allocation of individual reference fat contents
- The quota allocation year was equal to the calendar year (from 01 January to 31 December, the same year)
- Recently the transfers of quotas are not allowed (there are some exceptions).

Milk Product Council (MPC)

- Located in Budapest. Members are milk producer holdings and processor undertakings.
- Staff of 22 full time employees. No national network, field officers, equipment, etc.
- Carrying out administrative controls only.
- Summary: MPC cannot fulfil the requirements of the EU concerning milk quota management.

Necessary changes made

- In April 2003 decision was made by the leadership of the Ministry concerning the responsible body for quota management
- ARDA has been pointed out as the responsible body of the milk quota management in Hungary with the assistance of Livestock Performance Testing Ltd.:
  - Using its professional knowledge and co-operation during on-the-spot controls,
  - Administrative development of documents.

Preparation of ARDA and LPT being in progress from last October on (see Figure 3).

- Taking over the database of the MPC (producers, purchasers, processors – recorded files of allocated quotas). This first step is under progress.
- Assessment of necessary developments and needs (skilled staff, equipment, IT background, etc.)

- Estimation of the costs of these developments and adjustments. Acquiescence of the estimated costs by the Ministry of Agriculture – placing the money at ARDA's disposal and executing the developments.

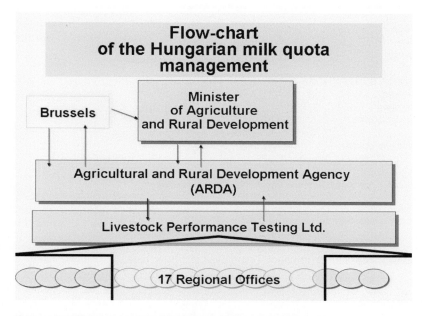

Figure 3. Flow-chart of the Hungarian milk quota management.

Application of quota system

1. Why the Agricultural and Rural Development Agency?
- ARDA is the official Paying Agency of Hungary
- Controlled by the Ministry of Agriculture and Rural Development (MARD)
- Skilled staff available, nationwide network available
- Headquarter in Budapest
- High level IT support (data processing) available
- IACS prescription.

2. Why the Livestock Performance Testing Ltd.?
- 100% owned by the Ministry of Agriculture and Rural Development (MARD)
- Skilled staff availableTraditionally good and regular contact with herd owners in terms of services (95 years) Headquarter in Gödöllő (about 30 km from Budapest to the North)
- 17 regional offices in 19 counties in Hungary (so national network already available)
- Accredited laboratory in GödöllőHigh level IT support (data processing).

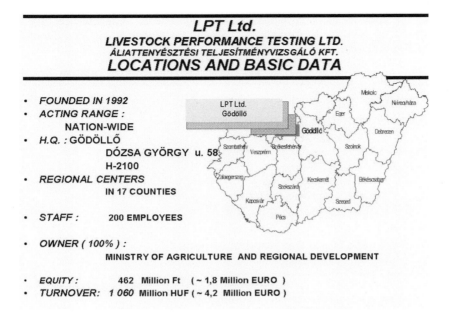

Figure 4. Location of Livestock Performance Testing Ltd and basic data.

The herds recorded by method „A" represents~ 3% of potential (~ 26,000) milk producing farms, with ~70% of total dairy cows produce ~ 83% of the yearly milk production (~1,600 billion kg).

The herds recorded by method „A" and „B" together represents: 5% of potential (~26,000) milk producing farms, with 72% of total dairy cows produce ~ 85% of the yearly milk production (~ 1,650 billion kg).

The herd size, the level of production and the proportion of the total milk production provides a distinguished position for the Livestock Performance Testing Ltd.

Table 5. Distribution of cow population according to type and milk recording methods (december 2003).

Total cow population	**337,000**
- Dairy cows	298,000
- Beef cows	39,000
Recorded dairy cows	**228,000**
- by method "A"	221,000
- by method "B"	7,000
Non-recorded dairy cows	**109,000**
(in ~ 25,000 small herds)	

Table 6. Distribution of herds recorded by methode „A'' – herd size (June 2004).

Cows/herd	Herds No.	Herds %	Cows No.	Cows %
= < 30	93	13.5	2,746	1.3
51 - 100	71	10.3	5,099	2.4
101 - 300	213	30.9	41,992	19.6
301 - 500	177	25.7	69,378	32.5
501 - 600	47	6.8	25,895	12.1
601 - 700	44	6.4	28,195	13.2
701 - 800	15	2.2	11,130	5.2
801 - 1000	15	2.2	13,427	6.3
1000 <	13	1.9	15,861	7.4
Total	688	100.0	213,723	100.0

Average (cows/herd) = 311

The LPT Ltd is *in contracted contact* with dairy farms *for basic* ("traditional" milk recording) *and surplus* (farm management assisting) services. The range of operation and services exist of:

a. *Basic* activity and related services:
- For the improvement of cattle population
  - o production control, collection of herd-book data,
  - o milk component tests,
  - o data processing, information service.
b. *Surplus* services besides the basic activity
- For producing better quality milk
  - o somatic cell count test,
  - o microbiological test,
  - o udder health services.
- For the improvement of herd structure and fertility
  - o evaluation system for reproduction.
- For the improvement of feeding efficiency
  - o urea test, feeding-monitor system,
  - o acetone test.
- To support breeding and production management economy program (farm management),
- For the promotion of economical operation of dairy farms - economic efficiency analysis (economy program)
- For the improvement of the structure of cattle population: full-scale collection of cow-insemination data (TER).
- To establish breeding operations and economic milk production on small dairy farms; Milk recording in small dairy herds:
  - o production control, collection of herd-data,
  - o milk component tests, SCC tests included,
  - o data processing, information service.
  - o row milk qualification.
  - o advisory for small-scale producers

Experiences with quota system

Data about buying and selling of quotas:
Buying up by state at November 2003 and April 2004
- Total quantity of quotas: 121 million liters
- Approx. 5 500 producers sold individual quota

Selling quotas by state at November 2003
- Total quantity of quota sold: 36 million liters
- Approx. 150 producers bought individual quota
- Recently new selling of quota is being in progress

Table 7. Quota and exchange.

| | Gov. Quota LEASING I. (in progress) – August 2004 | | | |
| | Owner | Quota to lease, | Price + VAT/litre | |
For lease:		kg	HUF	~ Euro
Delivery	?	7,000,000	1.00	0.40
Direct				
- A: (Quota had been lossed)	?	3,000,000	1.00	0.40
- B: (Never had quota)	?	120,000,000	1.00	0.40
Total	?	130,000,000	1.00	0.40

Are there any side-effects of quota system...?

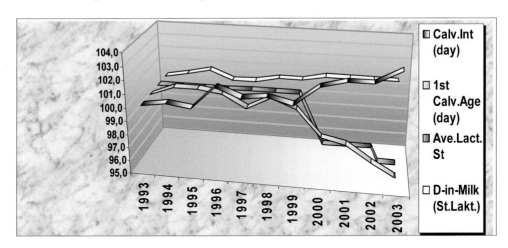

Figure 5. The changing balance between productive and reproductive traits together with increased milk yield (1993=100%).

- Less cows have to produce the same (in some countries: more?) amount of total milk.
- The balance between " the productive" and the "re-productive" traits seems to be broken!

The following perhaps "unexpected" tendencies can be signalled in the Hungarian cow population and probably also elsewhere.
- Increasing milk yield per cow;
- Decreasing age-at-first calving;
- Increasing calving interval;
- Decreasing lifetime (in terms of "average lactation stage" means: a decreasing proportion of cows complete more than their 3rd, 4th, 5th and more lactations);
- Decreasing lifetime production.
- Less calves born;
- Less roughages and more concentrates consumed.

The big question is: are these tendencies side effects of modern management and breeds or just temporary phases in a development traject of dairy farming and cattle breeding. Moreover: do these tendencies fit under a quota system? The answer may be: no!

These tendencies and their consequences will have to be considered in the future of dairy cattle farming in the EU!

Figure 6. Hungarian Grand Champion Cow 2005.

Structural and farm developments as consequence of the introduction of milk quota and suckler cow and beef premiums in Czech Republic

Oto Hanuš[1] and Jindřich Kvapilík[2]

[1]*Research Institute for Cattle Breeding - Agrovýzkum, Ltd., Rapotín, and National Referential Laboratory for Raw Milk, Vyzkumniku 267, 78813 Vikyrovice, Czech Republic*
[2]*Czech Moravia Breeders Association, a.s., Prague, 25209 Hradišťko pod Medníkem, Czech Republic*

Milk production and milk quota system in the Czech Republic

Similarly as in majority of European Union states, the milk production is traditional and important area in the Czech agrarian sector. That is reason, why relevant attention is put on milked cow keeping, on milk quota problems and on common organization of milk and milk product market.

The development of basic parameters of milk production

Since the change of political and social system in 1989 the long period recession has started in the cattle farming and especially in the milk production similarly as in the other agrarian sectors in the Czech Republic (Table 1). It has been caused by main reasons, which have been as follows:
- significant decreasing of economical support from state sources;
- increasing of input prices;
- belongings restitutions;
- privatization and restructuring process of agricultural enterprises;
- indebtedness of agrarian sector and its insolvency;
- increasing of call for milk and milk products.

Table 1. Parameters of milk production.

Parameter	unit	1989	1996	2000	2002	2003
Cattle in total	ths.	3 480	1 989	1 574	1 582	1 474
Dairy cows	ths.	1 248	657	515	477	460
Milk per cow	kg	4 101	4 429	5 412	5 890	5 929
Milk production	ths. tons	5 038	3 130	2 789	2 810	2 725
Market milk production	ths. tons	4 607	2 611	2 589	2 612	2 607
Milk marketability	%	91.4	83.4	92.8	93.0	95.7
Home milk consumption	ths. tons	3 172	1 966	2 083	2 129	2 143
Milk exports	ths. tons	1 435	798	689	662	795
Milk imports	ths. tons	x	144	193	248	289
Milk fat	%	4.00	4.02	4.03	3.98	3.97
Milk production for own domestic consumption	%	145	133	124	123	122
Farmer milk price[1]	€/100 kg	15.98	21.57	23.37	25.21	24.31

[1] quality class Q; 1 € = 32,00 CzC.

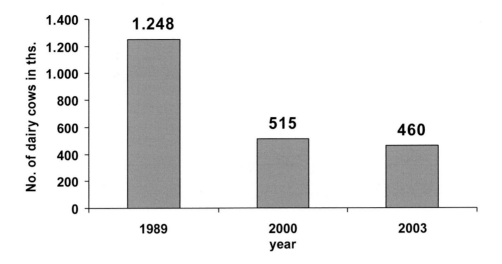

Figure 1. Number of dairy cows in years 1989-2003.

In the period 1989 – 2003 the counts of cattle in total were decreased from 3,48 to 1,47 mil. heads, it means to 42%. The same figures at the dairy cows were from 1,25 to 0,46 mil. heads, it is to 37% respectively (Figure 1). Milk production, deliveries of milk for the processing and home milk consumption decreased by 46, 44 and 32% in the same period. Since 1998 the year milk production and deliveries are almost stable, nevertheless it has not be reached the stopping of the annual decreasing of milked cow counts until now.

After significant decreasing (from 259.6 kg in 1989 to 191.9 kg per person in 1994) of home milk and milk product consumption per inhabitant it is necessary to evaluate the increasing during last years very positively (to 223.4 kg in 2003; Table 2). Similarly the average milk yield per dairy cow increased by 1,789 kg and 44%.

Table 2. Milk and milk products consumption per inhabitant (kg/year).

Parameter	1989	1994	2000	2001	2002	2003
Drinking milk	91.4	70.8	59.5	60.7	61.0	59.5
Butter	9.4	5.2	4.1	4.2	4.5	4.4
Cheeses in total	7.8	6.6	10.5	10.2	10.5	11.3
Curds	5.1	2.8	3.4	3.6	3.6	3.4
Other products	27.5	24.3	25.0	25.5	26.1	29.3
Milk tin	6.0	2.9	2.3	2.3	2.4	1.8
Total[1]	259.6	191.9	214.1	215.0	220.6	223.4

[1] in over calculation on milk, without butter.

The average milk yield of dairy cows in the Czech Republic reaches about 97% of average milk yield of dairy cows in the European Union (EU-15). The farmer prices of raw milk are lower by 10 or 20% in comparison to the relevant average prices in the European Union.

In opposition to the milk yield the trend of main parameters of cow reproduction performance shows the long period tendency to worsening. According to figures in Table 3 it is clear, that mentioned unfavourable trend has not been stopped in 2003. The conception rates of cows and heifers after first and all inseminations have been worsened. The insemination interval, service period (has been prolonged from 115 to 125 days during last five years) and between-calving time have been significantly prolonged. The count of birth and reared calves per one hundred cows has been decreased. Therefore, the improvement of dairy cow reproduction results is the main task of Czech dairy cow breeding together with keeping of favourable parameters of milk quality and improvement of economical results in next future.

Table 3. Conception rate after first insemination, service period (SP), calving-to-conception interval and calving interval.

Year	Conception rate after 1st insemination (%)			Days		
	cows	heifers	total	SP	calving-to-conception interval	calving interval
1998	47.0	63.5	51.7	80.1	115.2	400
2000	44.9	63.2	50.1	82.1	117.1	399
2001	43.9	62.1	49.1	84.7	120.3	400
2002	43.3	62.6	48.6	84.9	123.6	404
2003	42.7	62.2	48.4	86.3	124.6	408

A chosen results of milk yield recording

Over 95% of all dairy cows (Table 4) are involved in milk yield recording in the Czech Republic in last three years. It means, that majority of results, which are investigated in the framework of Czech milk recording has validity almost for whole population of milked cows (according to Kvapilík *et al.*, 2001, 2002, 2003).

Table 4. The milk yield recording (MYR) range in the Czech Republic.

Year	Cows[1]	Cows in the MYR		Method of MYR (% of cows)			
		total	%[2]	A_4	A_T	B	total
1998	561 600	524 780	93.4	91.4	8.5	0.1	100
2000	515 400	479 559	93.0	96.7	3.2	0.1	100
2001	483 500	471 370	97.5	98.2	1.8	0.0	100
2002	477 000	460 948	96.6	98.7	1.3	0.0	100
2003	460 000	443 750	96.5	98.8	1.2	0.0	100

[1] average year state;
[2] from total count of kept cows.

In 2003, the milk recording was performed in 2,075 enterprises and in 3,437 stables. It means, that there were included 1.66 stables per one agricultural enterprise (farm). There were kept on average 214 and 129 of dairy cows respectively on one enterprise and on one stable respectively. In total 92% of dairy cows were kept in 57.8% of enterprises with herd over one hundred and more dairy cows. Approximately 39% of dairy cows included in milk recording were kept in enterprises with cow numbers over 500 cows (in average 3.3 stables per enterprise). About 81% of all recorded cows were housed in stables with capacity one hundred and more dairy cows. Just 3.6% of cows with milk recording were kept in stables with capacity up to 50 animals. Only 0.2% of cows included in milk recording (Table 5) were housed in stables with capacity up to ten cows, which portion creates 7.7% of total count of stables.

Table 5. The rate of enterprises and stables in the milk yield recording according to number of kept cows (on February 3rd, 2004).

Number of cows	Enterprises (n=1,987)			Stables (n=3,096)	
	% of enterprises	stables in the enterprise	% of cows	% of stables	% of cows
from 1 to 10	5.5	1.0	0.1	7.7	0.2
from 11 to 50	21.2	1.0	2.6	17.0	3.4
from 51 to 100	15.5	1.0	5.3	27.6	15.6
from 101 to 300	30.3	1.4	25.5	35.2	43.0
from 301 to 500	15.5	2.0	27.4	10.0	26.6
over 500	12.0	3.3	39.1	2.5	11.2
Total	100.0	1.6	100.0	100.0	100.0

The population of milked cows has been created by c. 49% of Bohemian spotted cattle dairy cows (combined yield type), by 45% of Holstein dairy cows (milk yield type, including the crossing animals comes from transfer crossing) and by 6% of other breeds in 2003 (Table 6), as it is shown by distribution of cattle breed in milk yield recording system. The development of cow counts suggests on slow tendency to increasing of milk yield type of dairy cows, in particular in consequence of effort about improvement of milk production economical results in last years.

Table 6. The milk recording results according to dairy cattle breeds in 2003.

Breed	Cows[1]		Milk	Fat		Protein		First	Calving
	No.	%[2]	kg	%	kg	%	kg	calving[3]	interval
Bohemian spotted cattle	177,588	48.7	5,708	4.21	240	3.46	198	28/27	401
Holstein[4]	163,454	44.9	7,303	3.99	291	3.31	242	27/09	417
Montbeliarde	1,242	0.3	6,888	4.01	276	3.44	237	30/09	410
Ayrshire	300	0.1	6,008	4.18	251	3.37	203	30/00	421
Jersey	302	0.1	5,211	6.15	320	4.02	210	26/27	409
Others[5]	21,543	5.9	5,602	4.21	236	3.43	192	28/28	405

[1] number of cows with time of going to the end of standard lactation;
[2] from total number of ended standard lactations;
[3] age at the first calving (months/days);
[4] including the crossing animals from transfer crossing;
[5] other breeds and crossing animals.

The yield type of cows is expressed by milk yield, which is investigated by milk recording system. The difference between Holstein and Bohemian spotted cattle dairy cows in milk yield has been increased from 874 kg (per standard lactation) in 1998 up to 1,595 kg of milk in 2003 in milk yield type cow favour. It is comparable to differences, which are normally investigated between both yield types in dairy developed countries.

The evaluation of dairy cow counts and reached milk yields in dependence on main production areas in Czech Republic (Table 7) is very important from dairy cow keeping development point of view. In 2003, approximately 60% of milked cows were kept in highland (mountain and submountain) areas and 40% in lowland areas, similarly to situation in previous two years. The portion of dairy cows kept in highland areas has been decreased by 8.4% and in lowland areas by 14.6% in milk recording during last six years. Lower decreasing of cow counts under worse conditions means positive trend to exploitation of regions with higher ratio of permanent grassland by cattle rearing. The cow milk yield in lowland production areas is higher and is growing up quickly in comparison to highland areas. In 1998 the difference between both areas was 480 kg. In 2003 it was increased up to 595 kg of milk per cow. Relatively small differences between both areas are at milk protein content. Also at most other milk compositional parameters and properties are no significant differences. Significant differences were observed only at some cheese making technological properties (at no unambiguous tendencies) and at some figures of nitrogen phase of milk such as non protein nitrogen and urea ratio in non protein nitrogen (Hanuš *et al.*, 2004). However, these have not significant practice importance. The higher average age of the cows at the first calving is investigated in highland area and also slowly shorter between-calving period. Mentioned differences are in connection with higher contingent of Holstein breed cows in lowland area (c. 58%) and Bohemian spotted cattle breed cows in highland production area.

Table 7. The milk recording results in consideration of production areas (2003).

Production area	Dairy cows		Milk	Fat	Protein		First	Calving
	count	%	kg	%	%	kg	calving[1]	interval, days
Highland	217,457	59.7	6,181	4.17	3.40	210	28/19	406
Lowland	146,972	40.3	6,776	4.00	3.36	228	27/10	411

[1] age at first calving months /days.

Previously presented opinions about Czech raw milk production in context of decreasing

The total cow milk production is a little higher than milk consumption in the country nowadays. In total 2,810 mil. kg of milk was produced by 477 ths. of dairy cows in 2002. The count of dairy cows was reduced very sharply as mentioned previously. It is not good fact for CR agriculture. It was caused beside others mentioned reasons in particular by the milk product imports from abroad, which were very often subsidised by the governments of economical strong countries and by the decreasing of the home milk consumption as well. The other reasons were the custom conditions for milk and milk products, which were imbalanced (no equal or comparable respectively) in terms of export and import custom duty percentage and which were negotiated disadvantageously for CR in the framework of association agreements with the EU countries in mentioned period (Hanuš *et al.*, 2003a).

Raw cow milk quality testing

Raw milk quality, in terms of valid standards (EEC 92/46 and ČSN 57 0529, which is Czech national standard for quality of raw milk, which is delivered for heat processing into dairy plants), is investigated by the system of Czech central milk laboratories (now three laboratories). These are mostly accredited according to EN ISO/IEC 17025 standard for mentioned purposes by the Czech national accreditation authority (Czech Accreditation Institute, Prague). Laboratories take part in the different inter-laboratory comparisons about their milk analytical result accuracy regularly as well. The instruments are calibrated according to milk referential standard sample results as necessary. The results of the mentioned investigations are shown in Table 8 according to Roubal *et al.* (2003, 2004) during last seven (five periods) years. In general, the reached results are on relatively good level and comparable with results of majority of the dairy developed countries (Hanuš *et al.*, 2003b, 2003c). There is a trend for running improvement almost at all milk parameters too. In particular it is possible to observe such improvement at so called hygienic parameters. About 90% of raw cow milk deliveries were purchased in standard quality according to CR quality demands, which are in accordance with EEC 92/46 directive. In total 1,042,032 investigations of bulk milk samples for all relevant parameters were made about quality determination in 2003. Nevertheless, there are still reserves for the improvement, in particular at somatic cell count, total bacteria count and inhibitory substances occurrence frequency. Such are the next important tasks for Czech milk production system.

Table 8. Results of raw milk quality in CR.

Parameter	Unit	1997	1999	2001	2002	2003
Total bacteria count *	ths. CFU/ml	79.1	65.0	46.8	43.3	44.5
Somatic cell count *	ths./ml	237	248	259	259	261
Inhibitory substance occurrence	% of + cases	0.48	0.43	0.35	0.24	0.24
Freezing point *	°C	-0.5209	-0.5229	-0.5230	-0.5223	-0.5231
Psychrotrophic bacteria *	ths. CFU/ml	21.22	10.90	6.21	4.76	6.19
Coli bacteria *	CFU/ml	132	191	215	187	200
Termoresistent microorganisms *	CFU/ml	1275	1192	831	891	1174
Sporulate bacteria occurrence	% of cases	15.4	10.5	8.8	6.3	10.2
Protein *	g/100g	3.26	3.34	3.35	3.34	3.38
Fat *	g/100ml	4.27	4.24	4.19	4.14	4.13
Solids non fat *	g/100g	8.82	8.79	8.82	8.84	8.85

* The results are expressed by arithmetical means. According to Roubal *et al.*, 2003, 2004.

Milk quota system in the Czech Republic

Today state and development

A motivated demand of the Czech Republic on the milk quota for conditions of common market organization in the European Union was introduced 1997. It was done in accordance with Commission demand (Kvapilík & Vaněk, 2002). With regard to development of number of dairy cows and their milk yield, production, milk deliveries and home milk consumption, abroad milk and milk product market, milk production in European Union, targets of Czech agrarian sector and tasks of EU common agrarian policy, the Czech Republic demanded the allocation of milk quota at total amount 3,100 mil. kg.

The milk quota system was established by Governmental Regulation No. 445/2000 Sb. in the Czech Republic for quota year 2001/02. The preparation of milk producers, interventional organization (State Agricultural Interventional Fund) and milk processors (dairy plants) on using of regulations of common organizing of milk and milk products market in the European Union was aim of above mentioned standard. Therefore, the main directions from relevant legislation of EU were taken over into this regulation (quota year, penalty for exceeding individual quota, quota transfers, shortening for exceeding of reference amounts (quota), creation and distribution of national reserve, evidence system et cetera).

In consideration of motivated demand on milk quota under EU conditions (3,100 mil. kg) and actual production and milk deliveries and in accordance with Governmental Regulation (No. 445/2000 Sb.) and with other national legislation the individual (enterprise) production milk quota in amount of 2798,1 mil. litres of milk (c. 93% of demand 3,100 mil. kg) were distributed between producers (in total 3,602 enterprises) for quota year 2001/02. The milk quotas were shortened at 1,129 producers by 101 mil. litres from reason of quota insufficient gathering in mentioned quota year.

The 3,762 milk producers with identical amount of distributed national milk quota (2798,1 mil. litres of milk) were taken in the evidence of quota year 2002/03. 743.8 ths. litres of milk quota were connected with one producer in average. Such amount is balanced with keeping of 135 dairy cows per herd on one farm with allocated milk quota at average market production of 5,508 litres of milk per dairy cow 2003. The 2542.2 mil. litres of milk (90.9%) were gathered from allocated quota amount. The compensatory subsidy in amount 343.8 mil. Czech crowns were paid to milk producers during quota year.

The milk quota 2771,2 mil. litres of total amount were distributed between producers for quota year 2003/04. It is approximately equal to 2,850 kg of milk. Regardless of gathering of previously mentioned quota amount, the volume of national quota will be decreased onto level, which has been negotiated with EU, it means on 2,682,143 tons for quota year 2004/05. The linear decreasing of national quota for all producers is equal to 167.9 mil. kg approximately, it means about 5.9% of milk quota amount, which was allocated to producers 2003/04. No compensation was given to producers for economical loss, which by mentioned quota shortening them arose.

The Governmental Regulation No. 244/2004 Sb. from April 21st 2004 about specific condition determination for application of ration in milk and milk product branch in the framework common organizing of milk and milk product market was approved for the conditions of the Czech Republic membership in the European Union. This regulation harmonizes the national legislation which is linked with milk quotas with relevant EU regulations in terms of completion of problems of direct milk sale from farm, referential fat content, quota distribution from national reserve, quota transfers and others.

The estimation of influence of milk quota system and premiums on structural changes of dairy cow farming:

- During period since 2004 to 2010, it is possible to expect mild changes in the structure of agricultural enterprises and farms with milked cow keeping. The enterprises, which have improved significantly the housing and technology of dairy cow farming and which reach high level of work productivity, relevant milk quality and favourable economical results its production, will have interest about production expanding and about obtaining of larger amount of milk quota. On the contrary, it is possible to expect a limitation or cancelling of dairy cow keeping and milk quota sale at enterprises with obsolete technology of cow housing and milking and with unfavourable economical results of milk production. Such procedure will support evidently, up to certain level, the possibility of raw milk sale under more favourable economical conditions to abroad milk processors (in particular in border regions) during first two or three years of EU membership. The tentative estimation of some parameters of milked cow keeping up to 2010 is shown in Table 9.

Table 9. The estimation of selected parameters of milked cow keeping up to 2010.

Parameter	Unit	2003 (2004)	2010	Difference
Milk quota	tons	2 682 143	2 737 931[1]	+55 788
Milk per cow and year	kg	5 930	6 400	+470
Milk production enterprises	n	3 762	3 400	-362
Milk quota per enterprise	ths. of kg	713	805	+92
Dairy cows per enterprise	heads	135	130	-5

[1] including reserve for restructuralization (55 788 tons).

- The milk quota establishing and amount decreasing of national milk quota for conditions of the European Union in comparison to its distribution at national level for 2001/02 will have as consequence a limitation of milk production in some enterprises. The lower counts of dairy cows will not be replaced by the same cattle category in majority of enterprises owing to low quota of suckling cows. It is possible to expect the relevant expanding of bull fattening in limited count of cases. The receipts decreasing from milk production will be probably compensated by expanding of plant production on enterprise level. This development will be probably influenced by paying of majority of direct subsidies according to agricultural area. It is not possible to expect neither the significant changes at enterprise specialization nor expanding of enterprise in cattle rearing beside main working activity, which would be caused by milk quota system establishment.
- The establishment of milk quota system and direct payment according to area can influence the structural development of agriculture in terms milk production transfer into better natural and production conditions. The milk quota is in the fact national in the Czech Republic.
- In the enterprises which are focused on the dairy cow keeping (modern stables, high work productivity, good economical results), there is possible to expect the increasing of milk production efficiency, it means increasing of dairy cow numbers on area unit of agricultural land as well.
- For quota year 2002/03 the proportional part of reserve in volume of 100 mil. litres of milk was determined by Governmental Order No. 94/2002 Sb. (c. 3.6% of distributed quota). For the first quota year in the European Union (2004/05) a national reserve of milk quota is created in the comparable volume in the framework of over distribution of national quota on Euro quota. During next years the national reserve will be created mainly by delivery of part of transferred quota. Milk quota from reserve will be distributed two times per year. The contemporary producers (increasing of contemporary quota) and producers without previous allocation of any quota can request for quota allocation from reserve.
- The farmer purchase prices of milk about quality class Q are shown in Table 10 for years since 2000 to 2004. For the first two or three years of Czech Republic membership in European Union it is expected an approaching of milk farmer prices between CR and EU-15, it means their mild increasing in the Czech Republic. In the next period there will be expected the milk price development in accordance with reform principles of common agrarian policy of European Union and development of call and offer.

Table 10. Farmer purchase milk prices (class Q, Euro for 100 kg, 1 Euro = 32.00 Czech Crowns)[1].

Month	Year				
	2000	2001	2002	2003	2004
1.	23.19	23.56	25.22	25.03	24.38
2.	23.34	23.97	25.50	24.84	24.50
3.	23.38	24.03	25.47	24.72	24.53
4.	23.38	24.03	25.47	24.59	24.63
5.	23.38	24.16	25.38	24.28	24.69
6.	23.34	24.16	25.25	24.13	x
7.	23.34	24.16	25.16	24.03	x
8.	23.38	24.16	25.03	23.94	x
9.	23.38	24.16	25.00	23.91	x
10.	23.41	24.31	25.00	24.00	x
11.	23.47	24.63	25.00	24.06	x
12.	23.44	24.97	25.03	24.22	x
Average	23.37	24.19	25.21	24.31	24.54

[1]according to information of Czech Statistical Institute.

Milk quota and agricultural enterprise management

- Nowadays, the milk quota is in accordance with the today CR production. Therefore neither the conditions for milk production will be not changed in a dramatically way nor management of enterprises in relationship to milked cow keeping. A higher attention will be putted on better milk quality reaching, good health dairy cow state keeping, higher work productivity and better economical results and efficiency of raw milk production. However, at the fixed national milk quota and running increasing of milk consumption simultaneously in the country, in accordance with today trend, the real danger exists there, that Czech Republic will has to increase the milk product imports to cover its own demands in next future, which situation will be not so good for country economy balance.
- The milk quota system establishment and its current performance probably will not influence significantly neither today cattle breed structure nor breeder goals of individual breeds of milked cows in the country.
- The Governmental Regulation No. 244/2004 Sb. from April, 21st, 2004 makes possible a free transfer milk quota between producers. About establishment and performance of quota transfer by stock will be decided after evaluation of nowadays system. The quota price is determined by agreement between buyer and seller in dependence on call and offer at the milk quota transfer between producers. The prices of transferred quotas are not officially investigated and noted, there are no state evidence and therefore their heights are not known exactly. According to unofficial data, the milk quota price is approximately equal to farm purchase price of milk, it means between 15 and 30 Euro per 100 kg of milk. It is not speculated over short period milk quota transfers (leasing) according to Governmental Regulation No. 244/2004.
- The milk quota for EU conditions was negotiated in the amount 2 682 143 tons. For the quota year 2003/04 the quota 2 771 200 mil. litres (c. 2 850 000 tons) was allocated to farmers by the Governmental Regulation No. 445/2000 as mentioned already above. It means that distribution of Euro quota demanded a linear shortening already previously

distributed quota for all producers by c. 167 857 tons of milk, it is by 5.9%. The raw milk deliveries for processing at amount 2 607 ths. tons in year are in accordance with filling of Euro quota at 97.2%.

- The highest production costs of milked cow keeping are costs (at significant variability between enterprises) on own and bought feedstuffs (c. 40%), work costs (15%), energy, re-/deduction and reparation (11%), dairy cow herd replacement (9%) and insemination, veterinary service and drugs (from 4 to 6%). Therefore, the general goals for dairy cow farmers are to improve longevity, reproduction and health state of their dairy cows to reduce relevant mentioned costs and to reduce costs on feedstuffs at feedstuffs quality improvement as well.

Advisory service task in management of milk quota system

- The State Agricultural Interventional Fund is interventional agency, which is officially commissioned by relevant law to control, administer and perform the milk quota system in the Czech Republic. The special department of Ministry of Agriculture of the Czech Republic is commissioned authority to be responsible for national legislation, which is related to milk quota and for compatibility of home laws and regulations with regulations of the European Union. Commodity Professional Committee for Milk is the help organ for administration and performance of milk quota problems on national level in the country. Beside the representatives of State Agricultural Interventional Fund and Ministry of Agriculture the representatives of milk producers and their professional organizations are members of above mentioned committee as well. All these institutions and organizations are linked into advisory service system about problems of milk quotas. The pieces of experience of EU-15 experts, especially from Denmark, Germany and Austria, have been and are still used at milk quota system establishment. It has been and is done as by form of professional consultations as by form of professional lectures, training and schooling in the framework of European Union projects (Twinning, BABROC and others).
- The professional advisory service (schooling, training, seminars, publication in the professional press and so on) is necessary measure to adaptation of producers on conditions of common market organization, to optimal using of national quotas and to maximal implementation of direct payments.

Other different disunited opinions on milk quota system establishment and performance

Nevertheless, despite of above mentioned, it is necessary to mention, that there are not only supporters of milk quota system performance among representatives of Czech farmers and agricultural experts, but of course very often hard opponents (for example Kozel, 2001, 2003, 2004) too. They come mostly from larger agricultural enterprises in richer lowland or medium land areas, which have put higher costs into their breeder equipment and education during last time. Mentioned farmers reach regularly quite high milk yield of dairy cows. They are persuaded, that the milk quota system creates significant obstacles for their economical efficiency and growing, that the milk quota system is protected less able milk producers. Therefore they see the milk quota system as business brake with clear anti-market character and they are fighting politically against it in the country. Last but not least is their opinion that above mentioned fact is in strong discrepancy with clear declaration of the EU about necessity to overtake the economical level of USA by the EU countries. Although problem is controversial, they are right in terms of their own closer point of view.

Before the milk quota system establishment in the Czech Republic Kvapilík (2002) and Kvapilík & Vaněk (2002) discussed the EU and CR national quota limit proposals. According to author opinion the selected parameters of national milk quota, which were proposed by the EU Commission for CR, were significantly lower than comparable parameters reached in 1989, clearly lower than CR milk quota demands and mildly lower than actual state (2001) at mentioned commodity (after long period of reductions in cattle keeping numbers). Kvapilík & Střeleček (2003) have calculated in 2004, that volumes of quota limits, negotiated by the Czech Republic (at expressing on 100 ha of agriculture land), are on levels of 72.7% for milk quota, 83.7% for bulls and steers premium, 26.3% for suckler cows premium, 2.7% for ewes and goats premium, 49.0% for total livestock units and 64.7% and 13.6% for slaughter premiums an adult cattle and calves in comparison to EU-15 (Figure 2). In particular the quota for suckler cows is too low for necessary permanent keeping of grassland in good cultural and natural state. In this or similar continuity Havel (2001) independently mentioned, that first of all it was taken care of impartiality of milk quota system for all producers, processors and for member countries as well. This thought is questionable in terms of mentioned figures.

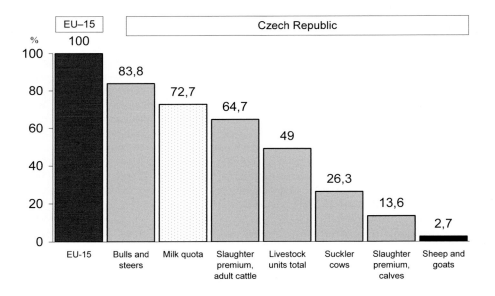

Figure 2. "Quotas" per 100 ha agric. land in EU-15 and CR (EU average = 100).

Holzner (2004) has modelled three differently modified economical scenarios for the eastern German, Czech and Estonian dairy system (farm) future development (up to 2010) to show how the milk producers in the old EU countries will be able to keep pace with those mentioned. His main question has been: „How competitive are the farms in the eastern accession lands now?" Author has predicted quite good development of milk production sector and profitable milk production for all three locations, however there will be probably a risk for over investments without first increasing profitability because of the high profit after

EU accession in case of Czech Republic and Estonia (those farmers will be encouraged to carry out large investments into milk production system according to author belief). This opinion is quite moot point as well.

Polish authors (Zachwieja & Knecht, 2002) are expecting for instance the significant impacts of milk quota system establishment and performance on structure of pulverized Polish sector of milk production in every case in next ten years. They see as main risk factor of Polish milk sector development the low level of allocated national milk quota.

Other animal science experts have evaluated the different related problems in last time too. They have estimated for example the possible influences of milk quota system on breeder dairy cattle targets and economical importance of milk traits for future breeding and dairy herd improvement in the Czech Republic (for instance Wolfová *et al.,* 2002). The authors have suggested in terms of future direction, that after milk quota (milk and fat amount) system establishment there will:

- increase the importance of milk protein from 35 to 41% of relative economical weight at Holstein (H) breed and from 41% to 44% at (C) Bohemian spotted cattle (two main milked groups of cows in the country);
- increase the relative importance of reproductive traits mildly;
- stay the milk yield traits as most important traits for selection;
- decrease the ratio of milk yield traits on total economical breed value (from 65 to 60% at H and from 56 to 54% at C);
- increase the ratio of beef yield traits on total economical breed value (from 21 to 24% at H and from 26 to 27% at C);
- increase the ratio of secondary functional traits (health, fertility, longevity) on total economical breed value (from 14 to 16% at H and from 18 to 19% at C).

Figure 3. Milk production is traditionally and important area in the Czech agrarian sector.

Conclusion

The relatively favourable agricultural enterprise and farm structure with dairy cow keeping, the milk yield and quality comparable with EU-15, the improved conditions of rearing and housing of dairy cows, the skilled workers in the animal production, the relevant management and ability for flexible producer reactions on changed market conditions, all mentioned facts create the presumptions to reaching of full gathering of negotiated national milk quota and to full ability to market competition for individual producers in comparison to milked cow farmers. It is clear of course, that in spite of all advantages of cattle keeping in the Czech Republic, the milk production under EU conditions will need significant effort, conscientious, responsible work and good co-operation of producers, processors, people in market and professional and state organizations as well.

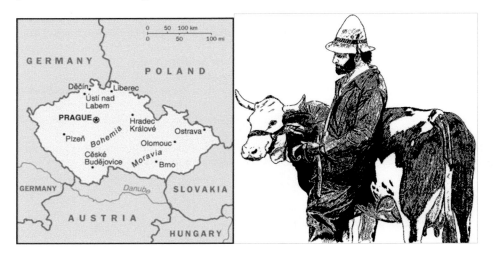

Figure 4. Map of Czech Republic.

Acknowledgments

This contribution processing for EAAP and FAO workshop and its presentation were supported by project solution budgets of MŠMT-ČR, MSM 2678846201 and INGO, LA 103.

References

ČSN 57 0529, 1993. Syrové kravské mléko pro mlékárenské ošetření a zpracování. Raw cow milk for dairy factory treatment and processing. Praha, Czech Republic.

ČSN EN ISO/IEC 17025, 2001. Všeobecné požadavky na způsobilost zkušebních a kalibračních laboratoří. The general demands on proficiency of testing and calibration laboratories. Praha, Czech Republic.

EEC 92/46, 1992. Milk and milk products quality.

Hanuš, O., M. Bjelka, V. Genčurová, R. Jedelská & J. Kopecký, 2003b. Vztahy průměrné velikosti stáda dojnic a některých kvalitativně-hygienických ukazatelů mléka. The relationships of the average sizes of the dairy cow herds and some qualitative-hygienical parameters. Výzkum v chovu skotu, 45, 3: 1-12

Hanuš, O., M. Bjelka, V. Genčurová, R. Jedelská & J. Kopecký, 2003c. Porovnáváme objektivně kvalitu syrového mléka? Do we do an objective comparison of raw milk quality? Náš Chov, 63, 6: 34-39.

Hanuš, O., V. Černý, J. Frelich, M. Bjelka, J. Pozdíšek, J. Nedělník & M. Vyletělová, 2004. Vlivy faremních podmínek modifikovaných nadmořskou výškou lokality na některé chemicko-složkové, zdravotní, mikrobiologické, fyzikální a technologické ukazatele kravského mléka a senzorické vlastnosti sýrů. The effects of farm conditions modified by over sea height of locality on some chemical-compositional, health, microbiological, physical and technological parameters of dairy cow milk and sensorial properties of cheeses. Acta Univ. Agric. et Silvic. Mendel. Brun., in press.

Hanuš, O., M. Klimeš, P. Hering, P. Roubal, P. Mihula & V. Genčurová, 2003a. Scheme and main results of milk production and testing in the Czech Republic. Proceedings: Pieno tyrimu sistemos dabartis ir perspektyvos integruojantis i Europo Sajunga. The current tendencies in milk testing and future prospects in the integration into European Union. Pieno tyrimai, Kaunas, 69-77.

Havel, F, 2001. Mléčné kvóty a náš vstup do EU. Milk quotas and our entry into European Union. www. AGRIS.CZ. Sborník, Den mléka 2001. Proceedings of Milk Day 2001, April 26th, 2001: 5-8.

Holzner, J., 2004. The newcomers are catching up. Agrifuture, Summer 2, 16-19.

Kozel, V., 2003. Dosáhnout bezkonkurenční konkurenceschopnosti. How to reach the highest ability for market competition. Šlechtitel, September: 15-20.

Kozel, V., 2001. Pohled na vývoj, stav a perspektivy chovu dojnic v Česku. The view on development, state and perspectives of dairy cow keeping in the Bohemia. Šlechtitel, 2: 6-13.

Kozel, V., 2004. Španělsko očima českých zemědělců. Hlavní španělské šoky. Spain by the Czech farmer eyes. The main Spain shocks. Šlechtitel, June: 22-30.

Kvapilík, J., 2002. Cattle farming in the Czech Republic before and after entry in the European Union. Report of Research Institute for Animal Production, Prague, Czech Republic, July: 22.

Kvapilík, J., J. Pytloun, P. Bucek, *et al.*, 2002. Ročenka 2001. Annual Report, 2001. Chov skotu v České republice. Cattle Breeding in the Czech Republic.ČMSCH, a.s., Praha, Czech Republic, 103pp.

Kvapilík, J., J. Pytloun, P. Bucek, *et al.*, 2003. Ročenka 2002. Annual Report, 2002. Chov skotu v České republice. Cattle Breeding in the Czech Republic.ČMSCH, a.s., Praha, Czech Republic, 110pp.

Kvapilík, J., J. Pytloun, P. Bucek, *et al.*, 2004. Ročenka 2003. Annual Report, 2003. Chov skotu v České republice. Cattle Breeding in the Czech Republic.ČMSCH, a.s., Praha, Czech Republic, 104pp.

Kvapilík, J. & F. Střeleček, 2003. Cattle and sheep quotas negotiated between the Czech Republic and the EU. Czech Journal of Animal Science, 48, 11: 487-498.

Kvapilík, J. & D. Vaněk, 2004. Možné dopady vstupu České republiky do Evropské unie na chov skotu. The possible impacts of Czech Republic entry into European Union on cattle breeding. Sborník, Den mléka 2004. Proceedings of Milk Day 2004, May, 5th, 2004, Czech Republic, 10-19.

Kvapilík, J. & D. Vaněk, 2002. Výroba mléka v ČR, v kandidátských zemích a v EU. The milk production in the Czech Republic, in the candidate countries and in the European Union. Náš Chov, 62, 7: 26-28.

Roubal, P., P. Kopunecz, *et al.,* 2004. Hodnocení jakosti syrového mléka v centrálních laboratořích České republiky v roce 2003. The evaluation of raw milk quality in central laboratories of the Czech Republic in 2003. Zpráva (Report of) VÚM Praha, Czech Republic, 20pp.

Wolfová, M., J. Přibyl & J. Wolf, 2002. Ekonomická důležitost znaků ve šlechtění skotu po zavedení mléčné kvóty. Economical importance of traits in the cattle breeding after milk quota system establishment. Náš Chov, 62, 4: 36-40.

Zachwieja, A. & D. Knecht, 2002. Příprava polského sektoru produkce mléka na společnou zemědělskou politiku. The preparation of Polish milk production sector on Common Agrarian Policy. Náš Chov, 62, 9: 25-26.

Farm management and extension needs under the EU milk quota system in Poland

Agata Malak-Rawlikowska

Warsaw Agricultural University, Department of Agricultural Economics and Farm Management, ul. Nowoursynowska 166, 02-787 Warsaw, Poland

Introduction

General overview of the Polish dairy sector

Poland is the significant dairy producer in Europe, with total milk production of about 11.5 mld. litres (Table 1), which places it on the 4-6 position in Europe (ex equo with The Netherlands and Italy). The dairy sector belongs to the most important sectors in Polish agriculture and food economy. It accounts (together with beef production) for 26% of Polish agricultural output value.

In 2002 there were about 875,000 of dairy farms from which about 50.2% - 440,000 were delivering to the market (direct sales and deliveries). In April 2004 there were 313 recognised purchasers of milk. Change of the economical system and drastic adjustment to the market conditions during the transition period in 90[5] caused about 40% decline of dairy herd. Slowly increase of the milk efficiency (19% during the period 1989-2002) couldn't have covered the herd decline what caused 28.5% cut in milk production.

Table 1. Production and deliveries of milk in Poland during the period 1989 – 2003.

	1989	1990	1994	1998	2000	2001	2002	2003
Number of dairy cows in 1000 heads	4,994	4,919	3,863	3,471	3,098	3,005	2,873	2,897
Milk efficiency in litres/cow/year	3,260	3,151	3,121	3,491	3,668	3,828	3,902	3,969
Milk production, mln l.	15,926	15,371	11,866	12,178	11,494	11,538	11,527	11,546
Deliveries mln l.	11,385	9,829	6,269	7,070	6,583	7,025	7,219	7,316
Share of deliveries in total milk production %	71.5	63.9	52.8	58.1	57.3	60.9	63.2	63.4

Source: Rynek mleka, Stan i perspektywy 1,21,22 IERiGŻ, MRiRW, B. Iwan SERiA 2002.

Organisation of the milk quota system in Poland

Polish organisation of the milk quota system is based on the European experiences and organisation orders in EU countries but, in regard to the special local requirements, it was not

---

[5] Freeing prices, abolition of production and consumption subsidies (which accounted about 60% of the price), and liberalisation of trade.

possible to implement one of the functioning in the EU systems. The final model of the quota regime in Poland is based on a regional organisation. The National Authority responsible for implementing and administrating the system is Agricultural Market Agency with support of its 16 regional branches (one in every Polish voivodenship). The general organisation scheme of the Polish MQS is presented in the Figure 1.

In the final pre-accession negotiations Poland received 8,964,017 t of milk quota (8,500,000 t wholesale quota and 464,017 t direct sales) and additionally 416.126 t for the restructuring reserve to be used as of 2005/2006. The first quota allocation has been made at the beginning of 2004, directly to producers, in respect of their deliveries during the reference year (01.04.2002 – 31.03.2003) and applications provided until the end of 2003.

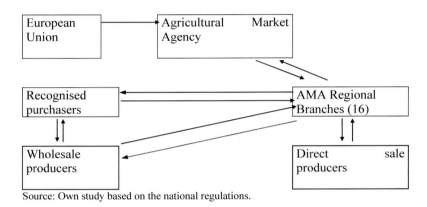

Source: Own study based on the national regulations.

Figure 1. Organisation scheme of the milk quota system in Poland.

Milk quotas for wholesale deliveries were allocated to the producers on the level of their milk deliveries during the reference year, so the allocation ratio amounted 1 (deliveries during the reference year accounted for 87.84% of the national wholesale quota). Regarding the direct sale quota, it was assigned to the producers after reduction by the 0.64 ratio. The reason was that the quota amount applied by the producers was higher than deliveries declared during the reference year and over twice as much as the national direct sale quota.

Referenced quantities have been specified in the Decree of Minister of Agriculture and Rural Development from 04.02.2004 (Dz.U. Nr 18, poz. 175), at the following levels:

National allocated milk reference quantity for:		National reserve of the referenced quantity for:	
- Wholesale producers	7.466.496.219 kg	- Wholesale producers	1.033.503.781 kg
- Direct sale producers	458.189.622 kg	- Direct sale producers	5.827.378 kg
Total:	7.924.685.841 kg	Total:	1.039.331.159 kg

In the general system AMA is responsible for keeping registers of dairy purchasers, directs sale producers and wholesale producers, for collecting and registering all applications for quota from the national reserve and for quota transfer. AMA is also monitoring the total amount of milk purchased by dairies, direct sale deliveries, all milk quota transfers, utilisation of national referenced quantity, levy payments and all the movements in the area of producers (i.e. changing purchasers) and purchasers (i.e. mergers).

In the system there are also involved the recognised purchasers, which play a role of kind of mediator between the wholesale producers and AMA. They are responsible for monthly and yearly reporting both to the wholesale producers and to AMA RB (regional branches). In case of the direct sale deliverers – they are under a direct registration of AMA RB.

Transfer of quota

Regarding the milk quota trade possibilities in Poland there are allowed the following ways of transfer: permanent transfer (to buy/sell/donation), temporary transfer by leasing (in/out) and conversion of the wholesale/direct sale quota (permanently or temporary).

- Permanent transfer agreements have to be delivered to AMA before the end of February each milk quota year. Regarding the regional organisation of the system, permanent transfer of quota is allowed only between the producers who have their farm in the same region of RB AMA. In case of the every permanent transfer the transferred amount of milk quota is being cut by 5% for the national reserve.
- Temporary contracts – leasing of quota, have to be delivered to AMA before 31st of January ongoing milk quota year. The leasing contract is valid until the end of the current milk quota year so 31st March. Leasing transfers are allowed only within the wholesale producers delivering to the same dairy purchaser and with farms located in the same region of RB AMA. The leasing transfers between the direct sales producers are to be made between the farms producing in the same region of RB AMA.

The milk production right can be leased out maximum for three following milk quota years. Besides the quota owner has to be conscious that if he leases out more than 30% of his quota (it will mean that he will deliver less that 70% of his quota) the unused by him part will be taken over to the national reserve.

Structural changes due to EU milk quota system

Structure of dairy sector in Poland

Dairy sector in Poland characterises rather high level of diffusion, not comparable with any other "old" EU Member State. Average dairy heard on a farm in 2002 accounted 3.3 heads (when in the EU already in 1997 was 24). Naturally when we take into account only the market delivering farms (ca 50% of all) we would receive the better average, but generally it is obvious that for polish dairy farms it is still long way to obtain the average level of concentration and market sale ratio characteristic for the EU.

During the transition process (since 1989), directly connected with radical changes of the economical conditions of production, average herd size increased only about 14% (2.9 in 1989 – 3.3 in 2002) when dairy herd declined ca 40% (from ca. 5 mln 1989 to 3 mln in 2001) and milk production also about 30% (from 16.4 mln t in 1989 to 11.5 mln t 2002).

Structure of dairy herd in Poland is illustrated in the Table 2, where can be observed that still about 64% of dairy cows are being kept in farms with 1 - 9 cows. The positive change was, that during the last 6 years, the total amount of cows in a group of "10 - 99 cows" has increased from about 9% of total herd in 1996 to 30% in 2002. The most of the cow population in that group lives in medium size dairy farms (10 - 29 cows), which account for 26% of total cowherd in Poland. Regarding the number of dairy farms it can be summarised that the biggest decline (37%) was observed at the group of small farms but still those producers are accounting for 93% of farm structure in Poland

Table 2. Number of dairy cows and number of dairy farms according to the farming size in 1996 and 2002.

Herd size in heads	Number of dairy farms in 1000				Dairy herd [%]	
	1996	Index %	2002	Index %	1996	2002
1 - 9	1287	98.40	818.9	93.60	85.7	63.9
10 - 29	20.4	1.56	52.5	6.00	7.5	26.1
30 - 49	0.428	0.04	2.25	0.26	0.5	2.7
Powyżej 50	1.245	0.09	1.35	0.14	6.8	7.3
Total amount	1309	100.00	875	100.00	100.0	100.0

Source: Powszechny spis rolny 1996, 2002 GUS, Warszawa, za Rynek Mleka nr. 25, X 2003r.

Generally it can be said that the concentration process in examined period was explained by moving the cowherd from the small farms to middle class farms – 10-49 cows, which produced in 2002 about 29% of milk in Poland (33% of total purchased deliveries). Therefore that group of producers will have the significant influence on a dairy market situation in the next years.

Effects of milk quota system on concentration process and structural development in European countries

In this report author uses such indicators as herd size, number of dairy farms and milk production per dairy farm to measure the structural changes. These indicators are also used by the European Commission and other authors (Oskam, 1991; Burrel, 1989; Runowki, 1994).

During the research, which purpose was to find the effects of the imposition of quotas on milk production concentration process, it was concluded that implementation of the milk supply limits affected the structural development and had slowed down the concentration process in the EU.

Limiting the market (free) increase of the milk production and from the other side expensive investments necessary to obtain an additional production limits made development of dairy farms and also market competition more difficult. In all the researched countries (Germany, The Netherlands, France, Denmark) and also in other countries (Switzerland, Norway, Japan, Austria) analysed in other publications, there was observed the significant acceleration in decline of numbers of smaller producers and reduction of cow population (Dillen, 1989; IDF, 1994, 1996; Oskam, 1991; Runowski, 1994). The milk yield incensement from dairy cow couldn't have covered the herd and farm reduction therefore increase rate of milk output from one farm was slowed down (comparing to observed in previous years). Figure 2 illustrate the evolution rate of milk production from one farm in Germany. Almost the same trends are observed in the other researched countries (the Netherlands, France, Denmark).

Figure 2 clearly illustrates deceleration of the milk production growth per one farm after 1984, even intensified in 1987 (introduction of fat correction of deliveries). Starting from 1990 the concentration rate again increased, mainly stimulated by the quota trade liberalisation (introduction of leasing and more liberal rules of transfers) and breaking the diminishing trend of dairy herd. This observation is also supported by the Figure 3, illustrating the dairy herd changes in the researched countries. There the significant decrease of dairy cow population appears after 1984.

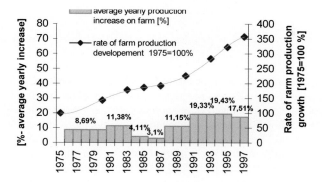

Figure 2. Rate of milk production concentration process in Germany (without ex-GDR), 1975-1997 (1975=100%).

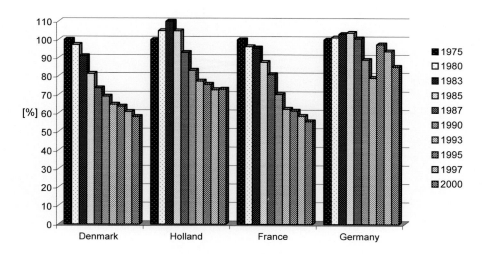

Figure 3. Changes in the dairy cow herd, 1975-2000 (1975=100%).

It was found (Burrel 1989; Oskam 1991) that with both purchase and lease transfers of quota, quota moves usually from the high marginal cost producers to low marginal cost producers. Purchase of quota was easier for farms with low production costs (usually bigger – scale effect) so before the leasing was introduced, the small farms rather tried to sell against to buy quota. That situation together with community (and national) cessation schemes accelerated the process of decrease in number of small producers. The introduction of leasing at the end of eighties gave to the smaller producers, with fewer assets a great possibility to develop their production. Since that time increased quota mobility had been additionally stimulating the milk production development at the farm level. (That efficiency development was also obtained by the milk yield increase and cost reduction per unit– i.e. reduction feed purchases, labour costs – like it was observed i.e. in France, Netherlands, England and Germany – Halliday, 1988).

Effects of milk quota system on concentration process and structural development in Poland

Accession to EU dairy market, especially execution of the EU standards and implementation of the CAP instruments, will accelerate development process of dairy production in Poland. It is estimated that this adjustment to a new reality (veterinary and sanitary norms, delivery requirements, quota allocation process) will stimulate producers to necessary investments and increase of production scale. But in addition these requirements will force a large group of small, ineffective producers to quit milk production, especially during the first years after accession (especially after the end of the transition period set at the end of 2006).

Introduction of milk quotas – a market supply limit, will made producers to increase efficiency of their production, first by improvement of milking facilities and herd reduction and then by the cost reduction. Therefore in the first years it is expected rather regular and intensive growth of milk yields and significant decrease in dairy population (see Figure 4 and 5).

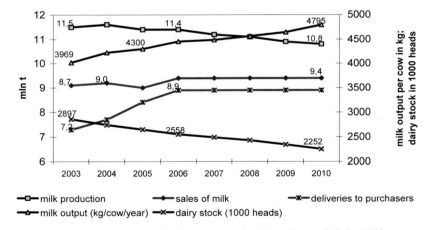

Source: Own calculations basing on EU Commission calculations, Seremaak-Bulge 2003.

Figure 4. Forecasts of the milk production indicators for Poland, 2003-2010.

Regarding the milk production forecasts (EU Commission, Seremak-Bulge, 2003) it is estimated that until 2010 the total milk production decrease to 10.8 mln t, mainly by diminishing self-deliveries and effect of milk quota on direct sales. Therefore that part of demand (moved from self-supply) will move to the market part of milk, what will stimulate a rise in milk deliveries to the purchasers. It is expected that development of milk deliveries will reach the level of wholesale quota in 2007 (deliveries in 2003 accounted for 82% of total wholesale quota). Concerning forecasted growth in milk consumption and taking under consideration all the abovementioned trends, it is possible that after 2008 internal demand for dairy products will exceed the available quota. Poland will become then the net importer of dairy products.

All mentioned before tendencies affect the farm concentration process, which – from rather intensive at the first 3-4 years after accession will probably slow down (observe Figure 5), as it was experienced in researched EU countries. Farm concentration is expected to speed up a little again at the end of this decade, after supposed effects of the new CAP reform (decrease in intervention prices, export subsidies etc.) and increasing quota mobility.

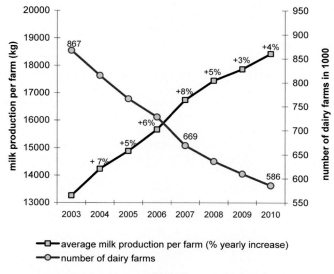

Source: Own calculations based on EUROSTAT Data –New Cronos Collection.

Figure 5. Concentration process forecasts for dairy farms in Poland (2003-2010).

Regarding influence of milk supply cut on farm specialisation level it was found out (Oskam, 1991; Dillen, 1989) that reaction to fixing output is the decrease in specialization. In the Netherlands and in England dairy farmers were switching to more beef and sheep breeding. Dillen (1989) observed that after introduction of milk quotas rate of declining type 41 farms (specialised dairy) was higher than for all dairy farms (when it used to increase before 1984). Regarding the share in dairy herd of type 43 (dairying, rearing and fattening combined), 71 (mixed livestock, mainly grazing) and 81 (field crops, grazing livestock combined) was no longer declining and sometimes even increasing. It could have been result of the expansion of other agricultural activities (rearing, fattening, arable crops, other grazing livestock) and/or decrease of a dairy herd on type 41 farms resulting in shift to 43, 71, 81 farm types. Concerning the polish specialised dairy farms it is rather difficult to estimate its number. Such data are not yet available.

Influence of quota mobility on structural development and regional specialisation

Possibilities of quota transfers and their administrative restrictions can be very important for the structural development. As it was proved in research studies (Oskam, 1991) the more restricted are the milk quota transfers the less stimulating effects have for the concentration process and structural development of milk production. When quota is freely tradable it usually moves from the producers with high marginal costs to more effective ones (with low marginal costs). Experiences with free regional transfers confirm that, when made mobile, quotas move from regions with very profitable alternative land uses to regions with relatively poor profitable alternative land uses. Therefore increase in quota mobility in Poland would have a positive effect on the degree of specialization and so improve the overall structure and competitiveness of the agriculture sector.

In Poland there is observed an intensive polarisation concerning the milk production. There are 5 regions with large and increasing dairy production: Podlaskie, Mazowieckie, Warmińsko-Mazurskie, Wielkopolskie and Kujawsko-Pomorskie, producing together 65% of purchase deliveries in Poland. This regions characterise also high number of large herds – (over 10 cows) where lives ca 50-60% of total cow population in a district (average for Poland – 37%).

In regard to the Polish transfer system, quota will be tradable within the particular districts (regions) so it can inhibit transition of quota from the less specialised milk production regions to more specialised ones, and delay an efficiency and specialisation increase. It can also rise a problems when in the less effective regions there are only few purchasers with EU approval and effective farmers have to transport their milk to the other purchasers (higher costs, less satisfying milk price).

Regarding the leasing transfers in Poland there will be possible to lease the wholesale quota only between the producers delivering to the same dairy purchaser. Impact of this restriction on structural development will be rather negative. It will significantly diminish an access to quota of the producers from the same region. It will also limit alternatives for the most efficient producers for gaining an extra quota.

Therefore it can be expected that generally such organisation of the milk transfer system would disturb the regional polarisation and specialisation in the milk production.

National reserve of the referenced quantities

Polish national reserve set in Decree of Minister of Agriculture and Rural Development from 04.02.2004 takes 11.6% (1.039.331.159 kg) of the total national referenced quantity. Ministry of Agriculture is responsible for establishing the disposal rules and measures for the national reserve, which have to be described every year. According to the specify rules the reference quantity can be obtained from the national reserve by:
- producers who want to develop their milk production;
- new established farms, producers beginning milk production;
- producers in case of positively reconsidered court appeals regarding the first allocation of the milk quota.

Producers developing the milk production have to prove that they have delivered from 1.04.2003-31.03.2004 at least 10,000 kg more than during the reference year; delivering after 1.04.2004 at least 10 000kg more than during the year 1.04.2003-31.03.2004 or during the period 1.04.2003-31.03.2005 delivered 20,000 kg more than during the reference year.

Appropriate management of the national reserve of dairy quota is very important for the support of the structural development of a dairy sector. It should be used for special measures stimulating the concentration process especially because, as it was concluded before, the introduction of the milk quota system has slowed down the dynamics of this process in many countries.

Importance of the outgoer schemes. Outgoer schemes (buy-out premiums) have various structural effects on milk production in particular countries – more effective in ones (like France, Germany, Denmark, Belgium) and less effective in others (Ireland, Italy, UK and the Netherlands). It was generally concluded in the research (Oskam, 1991, Dillen, 1989) that for countries with poorly structured dairy farming sector, with large number of a small producers (like i.e. Poland), introduction of the outgoer scheme could be a very good way of speeding up the structural development in the first years of a quota system.

Average milk prices 1998-2004 and forecasts of milk price

After rather high increase in 2000, milk prices in Poland decreased again in 2002. In 2001 nominal milk prices level was 29.1% higher than in 1999, and amounted 20.67 euro/100 kg. In spite of that rather high, for the polish conditions, price was still 35% lower that average price paid to the producers in the EU-15. In Table 3 there are presented the nominal purchase milk prices in Poland 1998-2004. In 2002 increasing trend stopped both in the EU countries and in Poland. Average milk prices in the EU had fallen down by 5.4% (amounted 29.7 euro/100kg) and in Poland ca 8.6% (calculated in zloty, 13% in euro) to the level of 19.16 euro/100kg (71.80 zl/100kg). In 2003 milk price in Poland have risen a little but regarding the high euro exchange rates it was only about 16 euro/100kg (EU – average 27.6 euro/100kg).

Table 3. Purchase milk prices (nominal) in Poland, 1998-2004, zł/100 litres.

	1998	1999	2000	2001	2002	2003	2004
Purchase milk price, yearly average per 100 l	60.69	60.84	78.35	78.48	71.80	71.51	78-83*

* Forecasts of IRGiŻ; Source: Rynek Mleka, Stan I Perspektywy 25/2003

The most important reason for the large differences between the Polish and EU milk prices are disparities in the production costs and different levels of market intervention on dairy market. Lower milk production costs were mainly affected by low labour costs (high unemployment), rather extensive character of dairy production and lower than in the EU quality requirements of the dairy purchasers. Moreover the significant influence on the prices has an intensive fluctuation of the euro exchange rates.

Milk purchase prices in Poland are competitive comparing the every EU country (the smallest difference in prices, 20-28%, comparing the Great Britain, Belgium and Ireland). In relation to new accessing EU countries, Poland has higher prices than Lithuania and Latvia (20-30%), which are therefore the greatest competitors for Polish products on the EU market. In the other new EU member countries milk prices are on the same or higher level than in Poland.

The new CAP reform (Luxembourg 06.2003), which starts 1st July 2004, is expected to have a large influence on EU milk price decrease. Reductions of intervention prices for butter (25%) and SMP (15%), together with cutting the intervention purchase limits are supposed to lower the milk price by 18-20% during first 4 years.

In Poland, the milk price forecasts are much more difficult and risky considering effects of transition periods and complying with the sanitary and veterinary requirements by dairy producers. After accession, expected increase in the export demand for polish milk products together with market stabilisation instruments, internal demand stimulation and export subsidies will probably stimulate the rise in the purchase milk prices. Factors enlarging the milk prices would be as well: expected cow herd reduction and milk production decrease, increase of production costs and improving milk quality. Nevertheless there will be visible still growing difference in the milk prices paid for EU quality milk and for the first class milk appropriate only for the internal market.

It is expected that polish milk prices will meet the level of the EU milk prices (which are thought to decrease) and until 2007 will rise some 10-20% (but rather won't be higher than 21-22 euro/100kg it is 90-95 zł - average purchase price and 23-24 euro/100 kg it is 110-120 zł for extra class). Further increase in the milk prices is rather not real in regard to expected EU prices decrease (effect of CAP reform), purchase power of the polish consumers and increasing competition on the EU internal market.

Milk production costs

Together with increasing milk prices it is expected that the milk production cost will also rise. As it was already said lower milk production costs are mainly affected by low labour costs (high unemployment) and rather extensive character of dairy production.

Table 4. Milk production costs and labour, grassland efficiency in Poland and EU countries in 2001 [euro/100kg milk].

Specification	Average for farms in Poland	EU average	Ireland*	Holland*
Total Income/100 kg	29.0	37.9	34.2	35.2
Including: milk output	25.0	32.1	29.4	30.4
Total direct costs	17.0	21.3	12.7	20.6
from which: Feeding	6.4	8.1	4.9	6.1
Machinery	3.3	5.0	2.2	6.5
Veterinary	1.1	1.6	1.2	1.8
Indirect costs	10.7	14.3	11.5	14.9
from which: Land	1.2	2.2	3.1	2.8
Labour	7.6	9.5	6.7	7.8
Capital	1.8	2.6	1.7	4.3
Milk quota costs	0.0	1.2	3.0	2.3
Total costs	27.7	36.8	27.3	37.6
Milk production per labour unit	67	155	-	-
Milk production t/ha fodder crops	6.2	12.4		
Milk production from grassland	3147.14	3061.0		
Number of cows/100 ha agricultural land	16	15		80
Grassland ha/cow	1.40	2.46		0.64

* data for year 2000. Source: Ziętara W. Runowki H , 2002 Materiały szkoleniowe FAPA; Ziętara W. 2002.

It was evaluated in the European Dairy Federation researches (EDF, 2000) that the total polish costs of the milk production was in 2000 ca 38% lower than the average in the EU countries. In Table 4 there are presented the milk production costs in chosen EU countries and in Poland in 2001. Analysing that cost specification it can be concluded that in 2001 average production costs in the researched group of farms were 25% lower than in the EU. It can be observed as well that the most important cost factors both in Poland and in the EU are feeding costs and labour, which together account for 50.5% in Poland and 47.8% in EU. Indirect costs share in cost structure is similar in Poland (38.6%) and in EU average (38.8%).

It should be mentioned that there were taken into account for Poland only market delivering, rather middle size farms. When we take into the consideration only the individual and smaller farms, the cost structure would be more charged by the indirect costs (ca 50%) and higher labour costs per unit.

Intensification of dairy farming

Rising milk efficiency by milk yield incensement, herd reduction and lowering the production costs will probably have an influence on continuation of rather extensive, land-taking milk production in Poland – see Table 4, (in 2001 – 16 cows/100 ha agricultural land; 6,2 t milk/ha of forage crops). It should be also emphasised that Poland has a good conditions for such production regarding the land availability and grassland area (1.40 ha/cow)[6]. Less extensive milk production (up to 6000 l/cow) based on the grassland feeding is more competitive that production based on intensive feeding systems (IERGiŻ, 1999-2000; Parzonko ,2003; German research). Therefore the economical management and unit cost calculations will have an important influence on the production scale and intensity of milk production. It can be supposed that limiting the milk market production together with other CAP implications will rather keep up extensive dairy production in Poland.

National support

Polish "Regulation on dairy market and milk products" from 6.09.2001 r. with following changes (Dz. U. w 129, poz. 1446) has introduced, since 1.01.2002, all the required instruments regulating the EU Common Dairy Market (except the international trade rules, which are laid down in separate regulation). Before introduction of that important decree Polish dairy market was supported by much smaller range of instruments (intervention purchases and storage, export subsidies, contingents and preferential credits for dairy since 1994). Regarding the additional supporting measures for polish dairy market in July 2002 there was implemented a direct payment for the best quality milk production ('extra class'), which aimed to encourage dairy producers to milk quality improvement. During 07.2002-04.2003 it amounted 4 gr/litr and later since 01.05.2003 – 7 gr/ litre. During the first year of its functioning purchases of milk in an extra class increased by 33%. After accessions that measure will not be continued.

The other national supporting instrument for dairy development is the preferential credits system introduced in 1994. Expenditures concerning that measure significantly increased from 6.1 mln Euro in 1995 to 14.8 mln Euro in 2003. During the negotiation process Poland got an approval for continuation of that measure after accession but on condition, that

---

[6] Main Statistical Office, Systematyka i charakterystyka gospodarstw rolnych, Warszawa 2003; Ziętara W. Organizacyjne i ekonomiczne aspekty produkcji mleka w przedsiębiorstwach rolniczych, Przegląd mleczarski 6/2002.

purposes financed by the national credits will be different than purposes supported by EU structural programmes.

Effect of milk quota on farm management

Milk quota prices for purchase and leasing

The milk quota values both for sale and leasing is being affected by various factors. Table 5 gives and overview of factors influencing the quota capitalization, which were found in the research studies (Oskam, 1991; Dillen, 1989; Burton, 1989; de Boer & Krijger, 1989; Stonehouse, 1990; Colman *et al.*, 1998).

The marginal cost and revenue basically determinate the quota prices, the first mainly affected by the feed and the second by the milk prices. But the future stream of net revenues will be fully capitalized into the milk quota value only when buyers are completely certain of it (Oskam, 1991).

It is rather difficult and risky to estimate the quota prices in the next years in Poland, regarding so many issues having influence on its capitalisation. Generally it can be expected at least at the first 3-4 years after introduction of the milk quota system the milk quota value will be rather low (purchase price estimated on the level of milk price, leasing – 10% of purchase price). It can be explained by taking into account the following factors:

- Polish quota is not tied to the land,
- there are rather high interest rates (costs for access to the capital),
- transfers between the producers (purchase) are deducted by 5%,
- quota assigned to wholesale producers haven't much cut their deliveries from referenced year (allocation ratio was 1),
- it is possible to apply for free quota from the national reserve,
- there is existing an "extra" reserve for structural development to be used as of 2006,
- total quota for deliveries is higher than recent wholesale deliveries (it is expected that deliveries will reach the level in 2007),
- there is assigned the transition period for levy payment and preparation of the milk quota system.

Moreover just after accession there has appeared a large supply trend by lots of small producers, who wanted to sell the quota and get some "free profit" form that cessation. The demand side was then even not yet built up, so they didn't succeed in selling the quota. That fact will also intend to quota price decrease.

As it was found, higher opportunity costs of production factors tend to lower quota values. When quota can be transferred without the land (i.e. Poland) the alternative use of land, labour and capital greatly influence the milk quota prices. In Poland standard gross margin from dairy production is one of the most competitive ones[7] in relation to the other land uses (so in that case the opportunity costs are lower) but in case of labour and capital it differ and depends on the production scale. Forecasted costs rise after accession together with additional cost component – "quota costs" and other mentioned before (chapter I) factors, would rather decrease competitiveness of Polish dairy farms, so it can also affect lesser level of the milk quota prices.

---

[7] Taking into account the market delivering producers with more than 8-10 cows, since where we can observe some profitability of milk production.

Table 5. Factors affecting the milk quota prices.

FACTORS	Effect on the milk quota value, when factor increase
Milk price	Increase
Opportunity costs of land, labour and capital goods	Decrease
Market demand for quotas	Increase
Supply of quotas	Decrease
Price of feed and other variable inputs	Decrease
Discount rate (interest rate)	Decrease
Expected duration of quota regulation	Increase·
General reduction of total quota	Increase
Levy rate	Decrease
Institutional arrangements	
• Transfer rules – quota cuts	Decrease
• Land attachment	Increase

Source: Oskam, 1991; Dillen, 1989; Stonehouse, 1990.

Probably, same as in case of milk prices, the milk quota prices will differ from one region to other and even (in case of leasing prices) between the particular dairy purchasers. It can be easily explained both by different profitability of dairy production among the regions, different costs factors and also by the transfer system chosen for Poland (regionally limited transfers, wholesale quota leasing between the producers delivering to the same purchaser).

Together with increasing quota turnover and rising demand for additional production limits average milk quota value will probably slightly develop, but in regard to all above mentioned factors and respecting effects of the CAP reform and deadline for milk quota arrangements at 2014, prices won't reach the high level.

Changes in management on the farm level

Adaptation of the milk quota system on the producer level will obviously involve significant changes in the farm management. First of all producers will have to introduce operational planning regarding the milk production level and its quota realisation. It means that they have to manage their production in order that not to deliver more that the quota allows and not less than 70% of their limit (in that case the quota is deducted by undelivered part). It also means that they will meet totally new problems in case of forecasted oversupply quota. For example producers would have to decide or to lease in the quota for covering planed oversupply or to reduce the further production by herd cut. Therefore there would be a need of economical calculations of marginal costs and expected profits to be obtained from both alternatives. Similar situation becomes in case of the quota trade possibilities and opportunity costs of quota leasing out (in comparison with production within the total limit). Above mentioned examples prove that on farm management will be much affected by the quota system.

On farm management will naturally by supported by monthly and yearly information received from the dairy purchaser (in case of wholesale producers) or AMA Regional Branches. That monthly information concerns (example for wholesale producer): amount of the milk quota with referenced fat content in ownership at the last day of finished month, monthly milk deliveries to the purchaser together with the average real fat content, monthly amount of fat corrected milk deliveries, amount of milk with referenced fat content delivered during current milk quota year until the end of month it concerns, amount of unused milk quota with referenced fat content to be used until the end of milk quota year. This set of information has to be delivered to the producer up to 25th day of the following month it concerns. On base of that information farmer has to manage his production and its development from one month to another. Therefore primarily dairy farmers will have to learn how to manage the information they receive but also how to search for an additional information concerning the milk quota system (i.e. transfer prices, contract forms, way of registration etc.).

It is rather clear that recognised purchasers and ODR (Agricultural Advisory Centres) will pay a significant role in advising to the wholesale producers in all kinds of decisions concerning quotas. For the direct sale deliverers that function will be taken mainly by ODR (regional Agricultural Advisory Centres). Well-organised information service would certainly improve effectiveness of farm management and general efficiency of the milk quota administration. Therefore it is very important for national authorities supplying an appropriate, competent advisory staff, organise traineeships and lectures for producers.

Organization of extension for the milk quota system

The milk quota system wouldn't be effective without an appropriate information service for all levels of its organisation. Therefore it was very important to establish an advisory program which would efficiently deliver necessary knowledge to producers, recognised purchasers and workers of all institutions involved in the system. Preparation and execution of training activities has taken a part of among others the Phare 2002 project, concerning implementation of the milk quota system in Poland. Generally the system of information spreading is organised in a cascade model as on the Figure 6 (Source: Agricultural Market Agency).

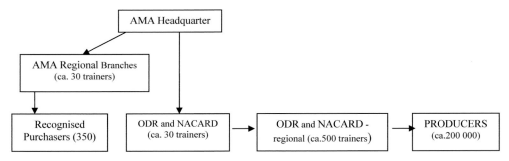

Figure 6. Organisation system of the advisory training program.

In the system of information flow there are involved both National Authorities (AMA with its Regional Branches) and Agricultural Advisory Centres (ODR) in co-operation with National Advisory Centres for Agriculture and Rural Development (NACARD). As it is visible on the scheme the information is transferred from the AMA Headquarter experts (also co-operating with external experts), through the middle chain units (AMA RB, ODR, NACARD) to the purchasers and dairy producers. Beneficiates of milk quota training will receive the summary of all important information in: brochures (for milk producers), Milk Quota System Manual, Milk Quota System Forms (lease form, heritage, conversion, movement, changing purchaser, direct supplier annual declaration), and EU and Polish regulations concerning the milk quota system. In regard to the project purposes, until the end of 2004, milk quota training will reach about 45% of dairy producers and all the recognised purchasers. System of milk quota supporting training will be continued for the next years.

Challenges for Polish dairy industry

Joining the EU structures creates great possibilities for the Polish dairy market as well for structural development as for efficiency and profitability increase. Nevertheless using these chances requires a lot of efforts and expenditures, and moreover needs very specific knowledge - how to use these possibilities for development. On this huge EU dairy market only the most efficient producers and dairies will succeed, being the ones which are competitive, modern and well managed. Therefore the biggest challenges for the Polish dairy sector in the next few years after accession will be to find the best usage of possibilities which the Common market gives us – that means improving milk quality, meeting the requirements set for the transition period, appropriate usage of financial devices and tools offered by EU for development and keeping up Polish competitive position in relation to the other EU member states.

Summary and conclusions

In the report the possible effects, which the milk quota system will have on the dairy sector in Poland, were presented. The following conclusions were obtained.

- In regard to the Polish milk quota allocation, it can be summarised that the model of the quota regime in Poland is based on a regional organisation. I is being managed by the Agricultural Market Agency with support of its 16 regional branches (one in every Polish regions). In the EU negotiation process Poland received 8.964.017 t of national milk quota (8.500.000 t wholesale quota and 464.017 t direct sales) and additionally 416.126 t for the restructuring reserve to be used as of 2005/2006.
- Concluding the structural changes due to milk quotas it is expected that accession to the EU dairy market, especially the execution of the EU standards and implementation of the CAP instruments, will accelerate the development process of dairy production in Poland. It is estimated that this adjustment to a new reality (including veterinary and sanitary norms, delivery requirements, quota allocation process) will stimulate producers to needed investments and to an increase of production scale. But in addition these requirements will force a large group of small, ineffective producers to quit milk production, especially during the first years after accession. These tendencies are expected to speed up the concentration process in the first 3-4 year after accession. But after that the concentration process will probably slightly slow down, and then speed up a little again at the end of this decade, as consequence of the supposed effects of the new CAP reform and of an increasing mobility of quota.
- In concern to the milk quota transfers it was assumed, that the more restricted the milk quota movement possibilities are, the less stimulating effects exist for the concentration process and structural development in milk production. In regard to the Polish transfer system, quota will be tradable within the particular districts (regions), so it can inhibit transition of quota from the less specialised milk production regions to the more specialised regions, and as such delay an increase in efficiency and specialisation in the dairy sector.
- After accession, the expected increase in the export demand for Polish milk products together with market stabilisation instruments, internal demand stimulation and export subsidies will probably stimulate a rise in the purchase milk prices. It is expected that

Polish milk prices will meet the level of the EU milk prices (which are thought to decrease) and will consequently rise some 10-20% until 2007. A further increase in milk price is not realistic in regard to the expected decrease in EU prices (as effect of CAP reform), the purchase power of the Polish consumers and the expected increasing competition on the EU internal market.

- Together with increasing milk prices it is expected that the milk production cost factors (recently about 25% lower) will also rise. It was analysed that the most important cost factors, both in Poland and in the EU, are feeding and labour costs, which together account for 50.5% of the total cost structure in Poland and for 47.8% in the EU.
- It is rather difficult and risky to estimate the quota prices for the next years in Poland, regarding so many issues having influence on its capitalisation. Generally it can be expected that at least in the first 3-4 years after introduction of the milk quota system the milk quota value will be rather low (purchase price estimated on the level of milk price; leasing – 10% of purchase price). Together with an increasing quota turnover and a rising demand for additional production quantities, the average milk quota value will probably slightly increase.
- Summarising the farm management changes due to the milk quota system it was obtained that the adaptation on producer level will obviously involve significant changes. First of all producers will have to introduce an operational planning regarding the milk production level and the realisation of the quota amount. Moreover, dairy farmers have to learn how to manage the information they receive, but also how to search for additional information concerning the milk quota system (i.e. transfer prices, contract forms, way of registration etc.). Therefore the establishment of an advisory program was very important. This program was planned to efficiently deliver the necessary knowledge to producers, recognised purchasers and workers of all institutions involved in the system.
- The advisory system for the purpose of milk quota management is organised in a "cascade" model of training's. It the system both the National Authorities (AMA with its Regional Branches) and the Agricultural Advisory Centres (ODR) are involved in co-operation with the National Advisory Centres for Agriculture and Rural Development (NACARD). It is expected that till the end of 2004, milk quota training's will reach about 45% of the dairy producers and all of the recognised purchasers in Poland.

Figure 7. Milk and milk products from Poland.

References

ARR, 2004. "System kwot mlecznych w Polsce" Podręcznik, Warszawa.

Boots, M., A.O. Lansink & J. Peerlings, 1997. "Efficiency loss due to distortions in Dutch milk quota trade", European Review of Dairy Economics, 24/1/1997.

Burton, M., 1989. "Changes in regional distribution of milk production" in A.M. Burrell Milk quotas in the European Community, CAB International, Oxford.

Burrel, A.M., 1989. Milk quotas in the European Community, CAB International, Oxford

Colman D., M.P. Burton, D.S. Rigby & J.R. Franks, 1998. "Economic evaluation of the UK Milk Quota System", Report, Ministry of Agriculture, Fisheries and Food, Welsh Office Agricultural Department, Scottish Office Agriculture, Department of Agriculture of Ireland.

Commission of the European Communities, 2002. Report on Milk Quotas, SEC(2002) 789, Brussels, 10.07.2002.

Court of Auditors, 2001. Special Report No 6/2001 on Milk Quotas, together with the Commission's replies, 2001/C 305/01.

Conway, A.G., 1989. "The exchange value of quota in the Republic of Ireland and some future issues for EC quota allocation" in A.M. Burrell, Milk quotas in the European Community, CAB International, Oxford.

Dawson, P.J., 1991. "The simple analytics of agriculture production quotas", Oxford Agrarian Studies 19, (2)

Dillen, M. & E. Tollens, 1989. "Milk quotas, their effects on agriculture in European Community" Volume 1, 2, University Leuven, Belgium

Domańska, E., 1998. Organizacja rynku mleka i produktów mleczarskich w UE, Problemy Integracji Rolnictwa, Unia Europejska – Polska,

De Boer, P. & A. Krijger, 1989. „The market for milk quotas in the Netherlands with special reference to the correlation between the price of land with quota and profit per ha in dairy farming" in A.M. BURREL, Milk quotas in European Community, CAB International, Oxford

Fedak M. & S. Stańko, 1999. Porównanie systemu regulacji rynku mleka w Polsce i Unii Europejskiej, Biuletyn Informacyjny ARR nr 9/1999r.

Gronowicz, M., 1995. Akademia Rolniczo-Techniczna w Olsztynie, Polityka państwa na nasyconym rynku mleka na przykładzie Wspólnoty Europejskiej (1), Przegląd Mleczarski, nr 2/1995r.

Gulbicka, B., 1997. Perspektywy produkcji mleka w świetle prognoz i zaleceń żywieniowych. Ierigż, Warszawa

Halliday, L.G., 1988. "Dairy farmers take stock: a study of milk producers reaction to milk quota in Devon", Journal of Rural Studies, 4, 3.

IDF, 1989. "Quota controls on milk supplies and supply management", Bulletin of the International Dairy Federation No. 245/1989, Brussels.

IDF, 1993."Twenty five years of European Community Dairy Policy", Bulletin of the International Dairy Federation No. 291/1993, Brussels.

IDF, 1996. "Market development for dairy quota", Bulletin of the International Dairy Federation No. 309/1996, Brussels.

IFCN World Wide, 2002. Dairy Report.

IFCN World Wide, 2003. Dairy Report.

Iwan, B., 2002. Kierunku przemian polskiego sektora mlecznego w świetle integracji i globalizacji rynku rolno-spożywczego, Roczniki Naukowe seria, Tom IV, zesz.3.

Krajewski, K., 1997. Organizacja rynku mleka w Unii Europejskiej, Problemy Integracji Rolnictwa, nr 2/ 1997r.

Krajewski, K., 2002. Warunki i potrzeby dostosowania systemu informacyjnego w polskim sektorze mleczarskim do wymogów systemu informacyjnego i regulacji rynku mleka w UE, Roczniki Naukowe seria, Tom IV, zesz.3.

Oskam A. & D.P. Speijers, 1991. "Quota mobility: the sale, lease and redistribution of milk production quotas in OECD countries", Report OECD, Paris, France

Rynek mleka – Stan i Perspektywy, market analyses, iergiż, 1999, 2000, 2001, 2002, 2003, 2004

Runowski, H., 1994. Koncentracja produkcji zwierzęcej", SGGW, Warszawa.

Seremek-Bulge, J., 2003. „Polski rynek mleka w pierwszych latach po akcesji do UE" Instytut Ekonomiki Rolnej i Gospodarki Żywnościowej (igrgiż), Warszawa

Seremek-Bulge, J., 2003. red. „Polskie Mleczarstwo" Raport o stanie branży i perspektywach jej rozwoju w poszerzonej UE, Związek Prywatnych Przetwórców Mleka (ZPPM)

Stonehouse, D.P., G.L. Brinkman, M.A. Mac Gregor & J. Tabi, 1990. "A methodology for evaluating maximum profitable bits for quota in the Ontario dairy industry" Department of Agricultural Economics and Business, Ontario Agricultural College, University of Guelph, Canada.

Świetlik, K., 2003. "Produkcja I ceny artykułów mleczarskich w Polsce I wybranych krajach",iergiż, Warszawa.

Tomkiewicz, E., 2000. Limitowanie produkcji w ustawodawstwie rolnym Wspólnoty Europejskiej, Wydawnictwo Naukowe Scholar, Warszawa.

Turvey, C., A. Weersink & M. Craig, 2002. "The value of dairy quota under a commercial export milk program", Working paper 02/12, Department of Agricultural Economics and Business, University of Guelph, Canada.

Ziętara, W. & H. Runowski, 2002. „Aktualna sytuacja I perspektywy rozwoju rynku mleka", SAPARD –FAPA, mrirw, Materiały szkoleniowe z zakresu nowoczesnego prowadzenia gospodarstwa rolniczego i finansów w kontekście integracji z UE.

Ziętara, W., 2002. „Organizacyjne i ekonomiczne aspekty produkcji mleka w przedsiębiorstwach rolniczych". Przegląd Mleczarski, 6/2002.

Influence of Poland's accession to the EU on milk production: Recent data

Andrzej Babuckowski

Agency for Restructuring and Modernisation of Agriculture, Warszawa, Poland

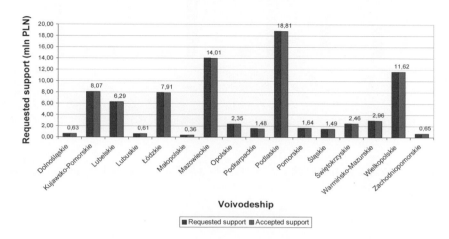

Figure 1. Financial support of dairy farms within SAPARD Programme according to region (per 3.11.05).

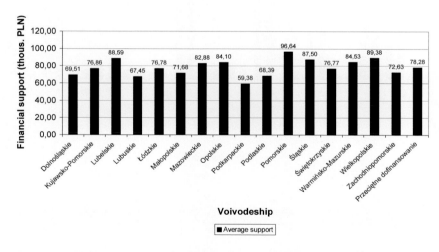

Figure 2. Average support per dairy farm within SAPARD Programme according to region (per 3.11.05).

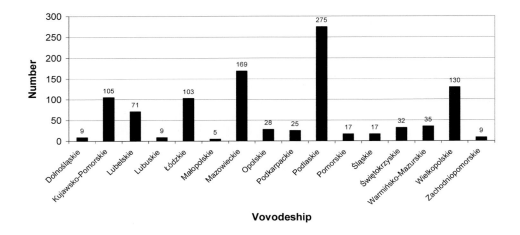

Figure 3. Number of applications from dairy farms accepted for financial support within SAPARD Programme according to region (per 3.11.05).

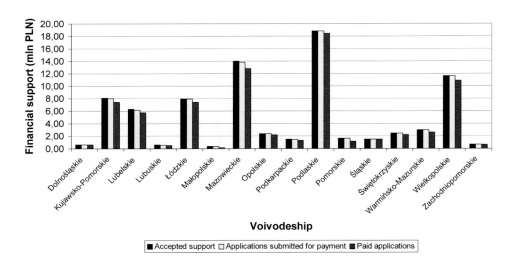

Figure 4. Realization of financial support for dairy farms within SAPARD Programme according to region (per 3.11.05).

Table 1. Polish data in 2004 – basic data.

No. of cows (thousand)	2 796
No. of dairy farms (thousand) – total amount	734
No. of dairy farms delivering milk to dairies	362
Average milk yield (kg/cow)	4 200
Milk production (mln tons)	11.8
Milk delivered to dairies (mln tons)	7.8
Average fat content in milk (%)	3.93
Milk purchasers	312
Milk sales to dairies (% of milk sales to domestic market)	87

Table 2. Factual results of the SAPARD Programme in dairy farms.

Indicators	Until 30.06.05	Total
No of completed applications	324	1034
Size of dairy herds (number of cows)	28	40
Average farm area (ha)	40.70	47.25
Value of investment	44 382 657,56	165 956 256,66
Value of support	21 491 556,81	81 163 028,41
Investments		
New buildings (number)	11	54
New buildings (area sqm)	5 266	21 812
Modernized buildings (number)	43	92
Modernized buildings (area sqm)	12 082	27 244
Developed buildings (number)	13	31
Developed buildings (area sqm)	1 877	64 433
Number of purchased equipment		
Milking machines (number)	82	268
Cooling equipment (number)	72	185
Pasture equipment (number)	6	16
Purchased animals (number)	454	1 681
Environmental measueres		
Capacity of containers for dung water	2 729	12 315
Number of containers for dung water	15	52
Capacity of containers for liquid manure	8 677	24 741
Number of containers for liquid manure	88	236
Area of manure plates (sqm)	11 221	40 465
Number of manure plates	71	231

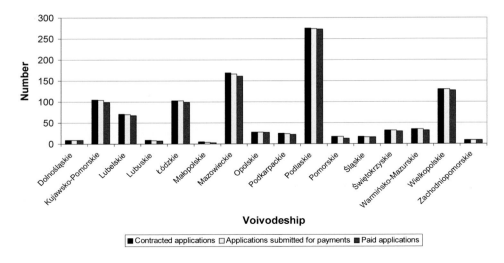

Figure 5. Realization of applications submitted by dairy farms within SAPARD Programme according to region (per 3.11.05).

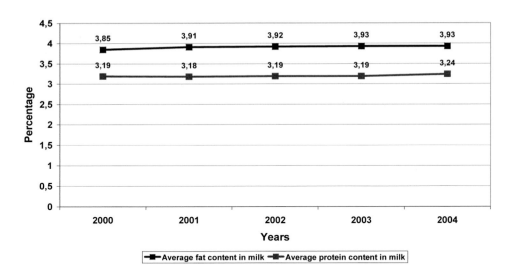

Figure 6. Changes in cows' milk composition.

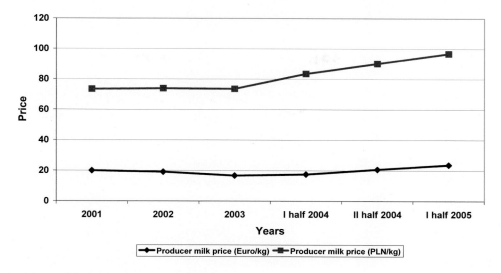

Figure 7. Changes in producer milk price.

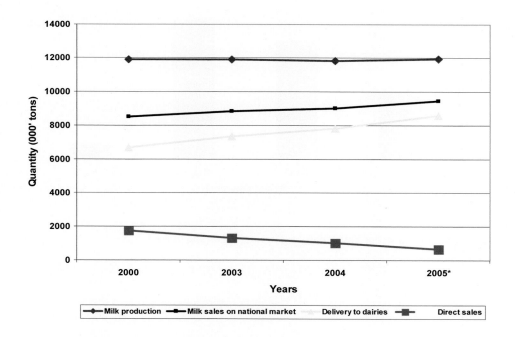

Figure 8. Production and delivery of milk to national market.

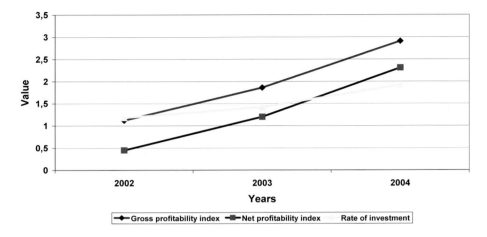

Figure 9. Economic results of the Polish dairy industry.

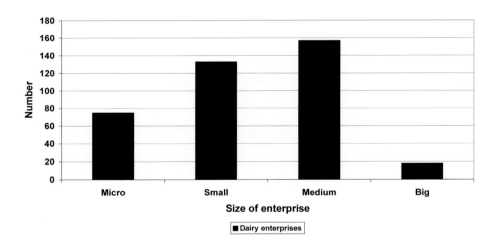

Figure 10. Distribution of Polish dairy enterprises according to their size.

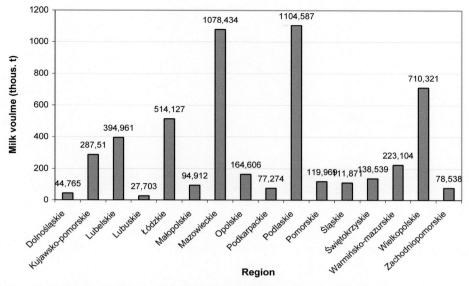

Figure 11. Volume of milk purchased by Polish dairies during first 7 months of year 2005.

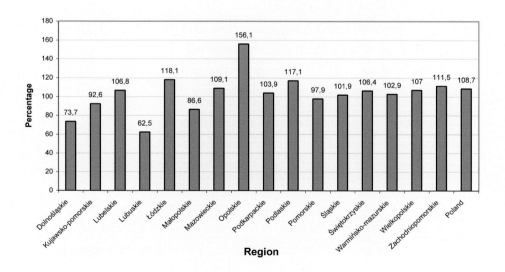

Figure 12. Change in milk deliveries to dairies (Jan-July 2005/Jan-July 2004).

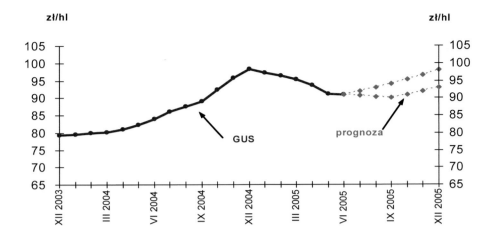

Figure 13. Milk price (without tax) in Poland and prognoses till December 2005.

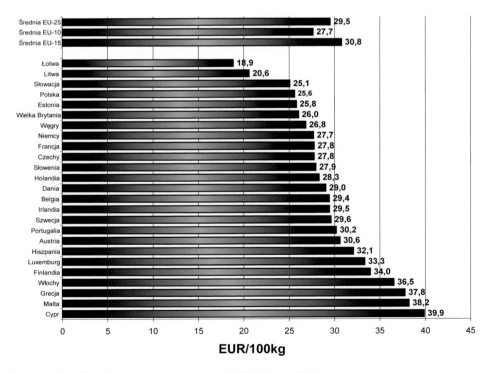

Figure 14. Milk price in member states of EU25 (Euro/100 kg).

Structural and farm development as a consequence of the milk quota introduction in Slovakia

Jan Huba[1], Štefan Mihina[1], Margita Štefanikova[2], Marian Zahumensky[2] and Jan Bročko[3]

[1] *Research Institute for Animal Production, Hlohovska 2, 949 92 Nitra, Slovakia*
[2] *Slovak Association of Milk Producers, Vystavna 4, 949 01 Nitra, Slovakia*
[3] *Agriculture Payment Agency, Bratislava, Slovakia*

Introduction

In recent years, the development of the dairy sector in Slovakia, like in other countries in transition, was affected by a marked decrease in number of dairy cows and subsequently a decrease in milk performance (Figure 1). Interesting is the fact that the decrease in number of livestock population was more intensive in our country than in other EU (15) countries, where milk production is regulated by the quota. During the transition period the reduction of dairy cows took place especially in the agricultural units with low milk yield herds. Low milk production efficiency was the consequence of an extremely high growth of milk production input costs and a stagnation in milk prices. Therefore it can be expected that in the first years after admission of Slovakia to the EU, the development in Slovak dairy sector will be identical with developments that happened in the „old" EU countries after the introduction of milk quota system. In these countries the expansion of milk production occurred before the quotation was in place. Irrespective of these facts, it is evident that mainly in the near future (we estimate after 2006) the quotation will influence not only the dairy but also other sectors in animal production in Slovakia, too.

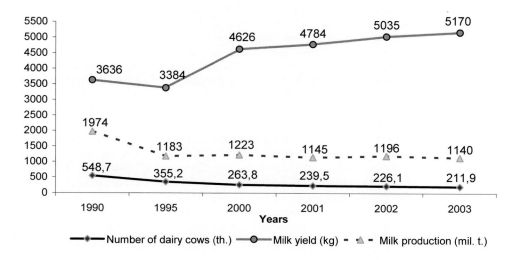

Figure 1. Development of dairy sector in Slovakia (1990 - 2003).

Expected changes in dairy sector in the next 5 years

Prediction of dairy cow population, milk production and milk yield

Prediction of the possible development till 2010 is complicated due to rapid changes on the European milk market. Therefore our prognoses are made for the sake of orientation only. Possible development regarding the number of dairy cows and farms, as well as other parameters are given in Table 1. We are expecting a heavy decrease in the number of dairy cows in 2004. Not because of quotation, but because poor financial situation of enterprises caused by postponed payment of EU subsidies at the end of the year, low quality of fodder crops, influenced by last year drought, and rising prices of slaughter cows. In terms of the year 2003, 1,140 mil. liters of milk was produced. During the first half of 2004 about 44% of the 2003 milk amount was produced, therefore we can expect a production of about 1,110 million liters of milk in 2004, which means that we will not exceed the milk quota.

Table 1. Prediction of dairy sector development (2004 – 2010).

	2004	2005	2006	2007	2008	2009	2010
Number of dairy cows (ths.)	207	203	196	195	190	185	182
Average milk yield (kg)	5,300	5,500	5,700	5,930	6,100	6,230	6,360
Total milk production (ths. t)	1,110	1,117	1,117	1,155	1,155	1,155	1,155
Number of large herds	800	780	760	720	700	670	660
Number of small herds	15,720	15,730	15,700	15,550	15,000	15,000	14,800

We expect the decrease in the number of dairy cows to slow down from 2005 onwards. In the next years the number of dairy cows should be directly connected to performance so that the total production would not exceed the national quota. There is a permission from EU for Slovakia to present the request to exceed the quota - national reference quantity – by 2% - from 1,113 to 1,140 millions kg in 2007. The condition for requesting additional milk quota will be that the current milk quota are fullfilled. We expect a larger increase in milk production between the two consecutive years compared with developments in the original 15 EU countries. The reason is the quite intensive Holsteinization during the last years (67% cows were inseminated with sperm from Holstein bulls in 2003), as well as the improvement of production conditions. The reason is also the improvement of competitiveness, since it will be necessary to achieve the average milk yield of the "old" EU countries with similar breeding systems (modern, expensive technologies).

As most of the Slovak farms can be considered as mixed farms (Table 2), it is complicated to extract the costs of the livestock production part from the total cost picture. Altogether livestock production contributes approximately 35% to income, indicating that milk production is of crucial importance from economical viewpoint. In future, a decrease of this share of animal production in agricultural income (more than 65% at present) is predicted, as the consequence of the change in supporting politics and the application of direct payments based on the area of land (abandonment of direct support of the animal production commodities).

Table 2. Farm structure in Slovakia (Farm Census, 2001).

	Number of large farms	Number of individual farms
Plant production	410	21,000
Cattle	117	5,234
Sheep, goats, horses	43	869
Pigs	53	6,993
Poultry	59	223
Mixed farms	872	32,530

Different production systems in Slovakia

It is complicated to estimate the development regarding the number of herds, too. Slovakia has two completely different farming systems in dairy cow production. Large herds (with more than 50 dairy cows) participate decisively in milk production. There are probably 178,000 dairy cows in these herds (85% out of total number). As we can see in Table 3 the highest number of farms (13,000) represents breeders with only one dairy cow. In the group of breeders with 2 – 50 cows (1,900 farms in total), prevail breeders with 2-10 cows. Herds with 1 - 50 cows which we will call small herds herein after, have the average size of 2.3 dairy cows per farm. During the last few years the development of large herds was accompanied by the decrease in the number of herds and the increase of the number of dairy cows / herd. We expect that setting up quotation will make this trend even more intensive and the concentration of dairy cows on farms will further increase. Besides the quotation there will be the pressure on the effectiveness of dairy farms. This process will probably cause a decrease in the number of workers on farm, which will be determined by the expected increase of labour costs. During the last ten years the number of workers on the farms decreased intensively mainly due to the modernization and change from tie to loose housing. This way the previous number of animal tenders (milkers, feeders) dropped to one third, resulting in a decrease of labour costs by more than 60%. The number of milkers decreased from 12 to 4 (including relieve milkers), and the number of feeders decreased from 6 to 2. We can expect that the increase of labour costs per worker will cause pressure on improving labour productivity on farms even further.

Table 3. Farm structure in Slovak dairy sector (Farm Census, Agricultural Payment Agency).

Number of cows/farm	Number of farms	Number of cows	Ratio of cows
1	13,000	13,000	6.1%
2 – 50	1,900	20,000	9.4%
51 >	820	178,000	84.5%
Total	15,720	211,000	100.0%

The small herds system is different and it is even more difficult to estimate its development. It is spread in specific regions, mainly in the Northern Slovakian regions Orava, Liptov and Spiš (see Figure 2). Milk production is determined mainly for subsistence farming. In general, these breeders have no quota for milk sale. In connection with the obligatory identification and registration of all cattle population, the recording of these farmers will improve. This information is provided by the Census of farms in 2001 in Slovakia. Most of these farmers have got no support in the form of subsides since 2003. Although after admission to EU they had the right to get direct payments, the predominant part of those farmers did not ask for these benefits. The main reason is the fear for bureaucracy. If the administration of direct payments would be easier and educational activities and extension would be improved, an increase in the number of small herds participating in the direct payment scheme may be expected (especially when the amount of direct payments comes to the level of the "old" EU members). However, for the future, it is more realistic to expect the opposite trend, because mostly elderly people are working in this system and younger people are not interested to succeed in this style of life.

Figure 2. The eight territorial administrative units (regions) in Slovakia.

Milk processing

As far as milk processing is concerned, this sector went through substantial restructuring after 1990. There are 108 of Approved Processors – with the Approval by Slovak State Veterinary and Food Administration. According to the Annual Amount of Processed Milk there are 31 enterprises processing more than 2 millions liters of milk, 27 enterprises processing from 500,000 to 2 million liters, and 50 subjects processing less than 500,000 liters of milk. The future development will be determined by the ability of milk processing companies success and then not only on the common market, but on the market of other European countries. Of course, all the exporters from Slovakia will have to obtain the respective certificates. Since foreign companies own the majority of the large dairy plants, their future will be determined by the situation not only in Slovakia but on foreign markets, too. We can expect further reduction in the number of dairy plants in connection with the milk quota system.

The influence of quota on other animal production sectors

As was already mentioned, the dairy sector has a strong position in Slovak agriculture (milk production represents almost 30% of agricultural revenues). The number of animals in other sectors is given in Table 4. Hitherto, development in the dairy sector (decreasing number of dairy cows and subsequently workers too) did not influence other sectors (animal number decreased in all of them, the number of goats fluctuate because there was a problem with the administrative recording of goats). The development in other sectors will be determined by the competitiveness of our breeders. The development of farm product prices will be decisive.

It has already led to the resolute decrease in the number of sows recently. In general, we can expect that pig and poultry production will not be influenced by the introduction of milk quota. We can expect a slight influence on ruminant breeding species. The decreasing number of dairy cows under the influence of quotation can lead to an increase in number of sheep, goats, suckler cows and / or horses. Remember, the decreasing number of dairy cows was compensated by the increasing number of suckler cows in many European countries. It seems that this trend will depend in Slovakia on the prices of weaning calves. If this price does not increase markedly, we can not predict great progress of this system. Similarly the development in sheep and goat breeding will be determined by economic efficiency, mainly by the prices of slaughter animals and of milk. The conditions of direct payment (keeping the land in suitable conditions for agricultural), and mainly the amount after adjustment to the original EU countries' level of payment can lead to the revitalization of ruminant production. Pig production seems the most promising branch from the view of the home market, since the demand is covered by imports.

Table 4. Livestock population in Slovakia (in ths. heads) (Statistical Office of the Slovak Republic).

	1990	1995	2004	EU Quota
Dairy cows	543.7	340.2	210.5	-
Suckler cows	5.0	15.0	30.5	28.0
Sows	179.9	160.5	91.2	-
Ewes	355.5	295.5	231.9	305.8
Goats	10.0	23.0	15.0	-
Poultry	16,477.8	13,382.4	17,200.0	-

At present, mixed farms prevail in Slovakia (Table 2). The experiences from the EU countries show the necessity of specialization. In the future we do not expect a growing number of mixed farm in spite of the influence of quota on the large farms. On the contrary, small farms with quota can be expected to replace (part of) the dairy cow herd by suckler cows or other ruminant species. Similarly, we can expect an increase of farm animals on the farms that are connected with agro tourism. This type of activity tends to increase especially the number of horses in Slovakia.

Current and expected milk price and costs

Milk prices significantly influence the situation in dairy industry. The rapid fall in the consumption of milk and milk products led to a surplus production in Slovakia after 1990. Now the average consumption per capita is about 165 kg (see Figure 3). The home consumption is about 70% of the total production.

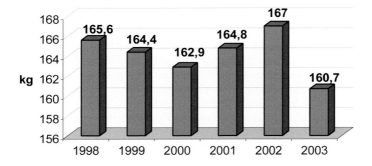

Figure 3. Milk and milk products consumption in Slovakia (kg/year).

The surplus production caused lower purchasing prices, which can not cover the production costs. Till 2003 this situation has been partially compensated by subsidies on a dairy cow basis. The support was 1 SKK per 1 kg of milk. In Table 5 the development of consumer purchasing prices in 2000 – 2004 are presented. In 2004 we have record a significant fall in consumption, which is caused mainly by the increase of food value added tax (from 6 to 19%). We also record the fall of the purchasing power of inhabitants, which is influenced by the liberalization of energy prices.

Table 5. Milk prices and milk consumption in the period 2000 – 2004 (Borecká, 2004; ATIS).

Year	Milk prices		Total consumption	Consumption per
	SKK/1kg	EUR/1kg	(ths. t)	capita[*] (kg)
2000	8.13	0.191	813,000	161.2
2001	8.55	0.198	825,550	163.4
2002	9.27	0.217	853,818	167.0
2003	8.95	0.216	810,305	160.7
2004	9.10	0.222	-	-

1 EUR ≈ 41,0 SKK (2004)

With regard to present turbulence on the European milk market, it is very complicated to predict the development of prices in the future. Before admission to the EU there was a vision that milk prices will rapidly rise to the prices of the original EU member countries. During the last months the EU prices fell very quickly and we were faced with phenomenon - the prices in the original countries draw to the price level of new member countries. This way the EU prices became equal to the prices in Australia, New Zealand and USA. This trend, connected with pressure on the liberalization of world trade will probably continue. In this connection we assume that the price will keep within the span 0.22 – 0.25 EUR/kg in future, with a tendency to rise in the next years (2005, 2006) and then the prices will fall again. We consider the development of prices as decisive factor from the view of further development of the dairy sector. A further decrease in the number of farms and dairy cows could be connected with it as presented in Table 1. The share of the largest cost factors in milk production in Slovakia are presented in Table 6.

Table 6. Overview of the main cost factors in milk production.

Item	Share (%)
Own feeds	24
Purchased feeds	13
Depreciation of animals	13
Other direct primary costs*	11
Other direct secondary costs**	12
Overhead costs	8
Labour costs	8
Depreciation of tangible investment property	6

*e.g. energy, insurance, veterinary costs
** e.g. transport, repairs of machines and equipment

National support

Slovakia has applied for so-called direct payment starting from the beginning of 2004. This payment and a national supplementary charge represent a level of 52.5% of direct payments of EU (15) in 2004.

In 2003, before the admission to EU the financial support for cattle and sheep production was as follows:

- dairy cows – up to 4,257 SKK (103 EUR) per animal
- slaughter bulls and oxen - to 2,250 SKK (54 EUR) per animal (min. carcass weight 285 kg, at export – the min. live weight 337kg)
- suckler cows - up to 8,500 SKK (205 EUR) in the group of land price 1 – 15 (less favourable areas) and to 5,000 SKK (120 EUR) in the group of land price 16 – 20
- sheep - in the group of land price 1 – 15 by breeding degree 650 –950 SKK(16 – 23 EUR) and in the group of land price 16 – 20, 350 – 500 SKK (8-12 EUR)
- sheep milk – at the sale 70 – 100 litres per ewe 6.5 SKK/litre (0.15 EUR) and at sale over 100 l per ewe 10 SKK/litre (0.25 EUR)
- genetic resources - 3,500 SKK (84 EUR) per cow of Pinzgau breed in protected herd and 3,000 SKK (72 EUR) per cow of Slovak Pied cattle in protected herd.
- The following support from national supplementary charge is intended for direct payment for cattle and sheep in 2004:
- suckler cows and pregnant heifers bred in suckling herd system (max. 50% dairy breed) – up to 4,000 SKK (100 EUR) per animal
- ewes and ewe hogs - up to 500 SKK (12.5 EUR) per animal
- genetic resources – Pinzgau cattle – to 3,500 SKK (75 EUR) per animal
- Tsigai and Valachian sheep to 1,000 SKK (24 EUR) per animal.

Effect of milk quota on farm management

Expected changes in management of farms and breeder's objectives

The quotation will probably influence the management of farms. In connection with the situation, described above (decrease in number of dairy cows before quotation), in the first years (till 2006 or 2007) the national quota probably will not be exceeded. The influence of quotation will be therefore more intensive in later years. The decisive factor will be the quota transfer and its price. We are not able to predict this today. This situation will mainly depend on the decisions of managers and the quota supply and demand situation.

During the first years the continuation of milk yield increase of dairy cows is expected, irrespective of quotation influence. It is necessary to achieve a level of performance which will be competitive. At the same time it will be necessary to improve significantly traits that are connected with fitness (fertility, longevity, good health of animals and milk gland). Some costs will be reduced this way. Significant will be also the emphasis on product quality for better realization of the products (to increase the income for the same amount of products). In connection with animal nutrition a pressure on quality forages can be expected, depending also on price development. There are still large reserves in this sphere in Slovakia. Mainly in the hilly and mountain regions it will be necessary to improve the utilization of grassland.

We can expect some changes in breeding work and in the representation of breeds. At present mainly the Holstein breed (67% inseminations) is used in breeding practice in Slovakia. The proportion of dual-purpose breeds is falling down (27% inseminations by Simmental - Fleckvieh, 2.4% Pinzgau breeds). The main reason for this development is the low prices of slaughter cattle to milk (4 : 1 in 2003). The increase in slaughter cattle price that has occurred recently can extend this ratio, which could lead to greater interest in dual-purpose breeds under the milk quotation system. Dual purpose breeds have usually also better values in fitness. Therefore it can be expected that some breeders will start using these breeds again.

Figure 4. Simmental cows as dual purpose breed again popular under the quota system?

We expect changes in breeding of individual breeds, too. The discussion concentrates mainly on the increase of the importance of fitness parameters and meat yield in selection indices. The ratio between proteins and fat changed from 3:1 to 5:1 in the production index (SPI) due to the quotation. This was calculated on the basis of economic weights determined under the milk quotation system. In the future we expect that traits connected with the state of health will be more pronounced in the selection indices.

Quotation, division of quota, transfer of quota and national reserve

Preparation of the legislation of milk quota is currently in its final stage. Therefore, we cannot answer yet all the questions connected with quota. In official documents National Milk Quota is called National Reference Quantity. National Milk Quota is divided into Individual Milk Quotas of milk producers, officially called Individual Reference Quantity. There are not any regional quotas in Slovakia. PPA – the Agricultural Payment Agency - administers the milk quota. It is responsible for request, approval, transfer and all other administrative procedures. Milk quota is linked to the cow. Transfer of quota is possible through purchase of the herd. It is not possible to buy or to rent milk quota.

From Quota Year 2003/2004 till 2007/2008 (31.3. 2008) there is a fixed national milk quota in the amount of 1 013 316 tons. It is divided into delivery quota 990 810 tons kg, and quota for direct sale 22,506 tons kg. The percentage between these 2 parts of quota makes 98% to 2%. Since April 1, 2007 - Slovak Republic is eligible to ask for an additional 27,472 tons, resulting in a national total amount of 1 040 788 tons kg.

We introduced the quota for the first time – last quota year. We started with the fixed National Milk Quota of 1,013,316 tons kg and fulfilled it by approx. 98%. (- 20 million kg).

System of allocation in 2004/2005

Individual Quota for 2004/2005 was allocated based on Individual Quota Allocation from quota year 2003/2004 that was approved by the Ministry of Agriculture and modified according to the fulfilment either by reduction if the quota was not fulfilled or in case the producers asked for an additional quota allocation.

Reduction of individual quota in 2004/2005

The system of quota reduction was applied according to the directive of the Ministry of Agriculture:

Quota Fulfillment	Quota Reduction*
90 – 100%	0%
80 – 90%	-50% from unfilled amount of quota
70 – 80%	-75% from unfilled amount of quota
Less than 70%	-100% from unfilled amount of quota

*Subjects who do not fulfil the quota are not eligible to ask for additional allocation

This way there is a total reduction amount from quota year 2003/2004 available for allocation in the amount of 23 million kilograms.

Additional allocation of individual quota in 2004/2005

Additional Allocation of Individual Milk Quota in quota year 2004/2005 can be based on a request of the milk producer to the Agricultural Payment Agency.

The Amount of Requested Additional Milk Quota Allocation this year is 40 millions kg.

Available amount from individual quota 2003/2004 reduction	23 thousand tons
- Minus National Reserve	2 thousand tons
- Minus New Allocation Requests = New Farmers	1 thousand tons
(max. 100.000 kg per farmer)	

After all reductions listed there is:

Available for additional allocation   **20 thousand tons**
Final additional allocation   **approx. 50% of the original request of 40 th. tons**

For instance, when the farmer requested additional milk allocation of 200,000 kg – he will receive about 100,000 kg.

Organisation of extension and extension needs

The Extension Service Network (ESN) has been created in Slovakia in 2000. This network included advisers, Extension centres, a Co-ordinating Centre and a Management board. The required number of extension providers has not been established. This is left to the role of the market and the level of educational qualification and experiences.

Extension Centres are situated in all regions of Slovakia. They are attached mainly to a larger independent organisation. They have direct contacts with farmers and other members of ESN. They help clients to crystallise the true nature of their problems and assist in identifying providers of extension services.

The Co-ordinating centre co-ordinates the activities of Extension centres, elaborates state policy priorities and registers qualified advisers and extension organisations. The Management board is responsible for the ESN strategy. The Board has 12 members from all parts of ESN.

Expected extension needs concerning quota and premiums - farmers want to know the following:

- if they can claim premiums
- what they have to do to obtain them (by administration)?
- how much will this administration cost?
- how to get to the sources from Sector operating programmes and funds on rural development?
- who will control drawing of financial means and how?
- what forms (administration) will be necessary for the quotation (sale to dairy processing companies, direct sale)
- how to manage the herds under quota
- searching for alternative production
- transfer of quota (possibilities, prices, legislation)
- levy for infringement and/or non - observance of the quota
- information about quota control.

The biggest challenge for the dairy industry in Slovakia in years ahead

The potential for dairy sector development in Slovakia is large in the view of the basic production volume. Compared to most European countries the advantage is the high concentration of dairy cows on farms. In many of them housing and technologies have been modernized during the past 10 years, and the world high quality gene pool has been used. Low labour costs and low price of land are advantages in comparison with other European countries. When the milk price reaches the price in the EU, Slovak farmers could produce milk more effectively compared with most European farmers (Figure 5). Expansion in this branch can be restricted by 3 factors. The first one is the quota; the second is low native milk and milk products consumption, and the third one is the quite complicated management of cooperative farms.

In case of continuing liberalization of trade and in situation without milk quotation a very dynamic development would start on many Slovak farms. In case of continued quotation, the development of the best farms will depend on the system of quota transfer and its price.

In view of milk processing, a high concentration of dairy cows on farms can be considered also as an advantage, together with improved quality parameters of milk obtained in modern milking parlours. Short transport distances are an advantage, too. The quite low total milk volume represents a disadvantage, and the quota system will restrict this volume in the future, too. This implies that milk processing capacities in dairy plants are not fully used, which could cause further reduction in the number of dairy plants in the future.

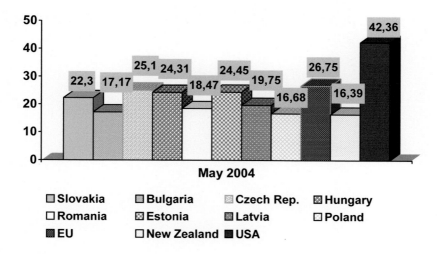

Figure 5. Producer prices of milk – country comparison (EUR/100 kg).

Quota and farm management in Lithuania

Egle Stonkute[1] and Violeta Juškiene[2]

[1] *Lithuanian Institute of Agrarian Economics, V Kudirkos 18, LT-03105 Vilnius, Lithuania*
[2] *Institute of Animal Science of LVA, R. Žebenkos 12, LT-82317 Baisogala, Radviliškio raj., Lithuania*

Introduction

Agriculture in Lithuania performs a significant social, ethno-cultural and environmental function. In the Strategy for Agricultural and Rural Development in Lithuania, the dairy sector has been acknowledged as a priority branch of agriculture (Figure 1). The transition from central planned economy to market economy has been very complicated and hard. A very high decrease in animal production has been recorded.

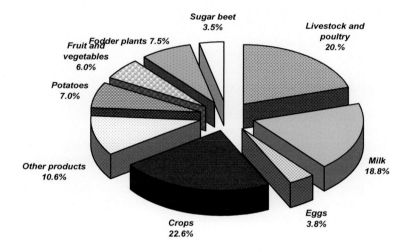

Figure 1. Structure of gross agricultural production in 2003.

One of the key goals following the integration of Lithuanian agriculture into the EU common market is the improvement of competitiveness (Figure 2). The future of the European market belongs to the modern, progressive agriculture based on the newest know-how technologies.

 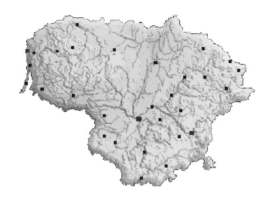

Figure 2. Europe with Lithuania pointed out.

Structural change due to EU-quota system and premiums

Today, three types of farms can be distinguished in Lithuania: farmers' farms, agricultural companies and individual family farms. The number of farmers' farms increased by 64% over the last year; agricultural companies work in a stable and promising tendency; the number of individual farms decreased. About 84% of all farmers' farms possess an area smaller than 10 hectares.

The economic position of farms depends on its size. The structural changes of farms is greatly influenced by their economic position and possibility to invest. Expected structural changes in dairy farming sector are presented in Table 1.

Table 1. Expected structural changes in dairy sector till 2008.

Item	Unit of measurement	2004	2005	2006	2007	2008
Dairy cows at the beginning of the year	1000 units	448.0	448.5	459.5	466.6	420.0
Dairy cows at the end of the year	1000 units	448.0	459.5	466.6	420.0	428.0
Average production per cow per year	kg	4,274	4,534	4,713	4,900	4,950

Source: LAEI calculations.

Mixed farming dominates in Lithuania, combining both crop and livestock production. Thus, the multi specialization is the trivial round in Lithuania. By contraries, CAP reform implementation is expected become an accelerator of farms specialization which would be very welcome in the country.

For a large part of producers, especially milk producer, farming is secondary source of income. It is not expected that full-time farming will growth significantly. Contrary, due to the whole agricultural policy, it is expected that the number of part-time farming will fall (Table 2).

Table 2. Farm structure.

	Number of farms	Land use, thou. ha	Average farm size, ha
Farmers' farms	39,249	871.5	22.20
Agricultural partnership	537	80.4	149.00
Individual farms	230,849	499.8	2.17
All farms	270,635	1,451.7	5.40

Agriculture in Lithuania is quiet regionalized: there are quiet important differences in soil quality, income levels in the regions. However, it is not planned, for a while, to set regional milk quotas.

Animal breeding in Lithuania is extensive. It reaches only 1 GV/ha during last three years. Thus, it is expected that the number of animals breeding per ha will growth during coming years.

There are 80 % of milk quotas distributed (95 % of quota for milk production and 61 % of quota for milk direct sales). In reality it is expected that milk quota reserve will be distributed among large farmers. There is no the sufficient potential of new milk producers which could apply for quota from the national reserves.

Milk prices are expected to growth (see Figure 3) from 2004. The prices of milk have grown 18 % from the beginning of the year 2004.

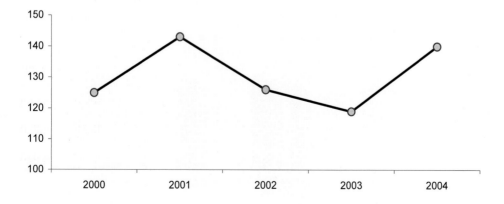

Figure 3. Milk prices 2000-2004.

National support

From 2004 direct support is implemented under SAP and CNDP schemes. There is no CNDP for milk in 2004. From 2004 only a part of payments for animals will be coupled, other part - integrated into single area payment.

In Lithuania milk producers used to be supported by payments for produced milk quantities (tones). Only in 2003 the payments per cows were added in order to provide supplementary support before the accession of the country to the European Union. In 2004 payment per tone are not planned because of the under preparedness for quota system administration and implementation of payments per tone. From 2005 it is planned to revert to payments per tone. The support for milk producer further was greater that it is expected to be under CAP reform implementation, with SAPS in mind which integrates a part of each sectors' financial envelopes into a single area payments.

Suckles cows support level (coupled part of support) was decreasing from 2001 and will start to increase from 2004 to reach almost the same level as in 2001. It is expected that suckles cows payments will continue to growth after 2006.

Bull special premium is introduced only under SAPS (CAP reform) implementation. These payments are expected to growth. Payments for ewes will be lower from 2004 that it used to be during 2001-2003. From 2004 the slaughter premium will be lower that it used to be during 2002-2003 (Table 3 and Figure 4).

Table 3. National subsidies in livestock sector 2001-2003 and SAPS implementation plans 2004-2006.

Premiums*	2001	2002	2003	2004	2005	2006
Dairy cows, EUR/animal			37.94			
For milk, EUR/t		4.84	9.47		6.95	6.66
Suckler cows, EUR/animal	173.77	158.42	154.65	144.81	157.84	170.3
Bulls, EUR/animal				147.71	156.39	169.14
Ewes, EUR/animal		24.04	18.25	11.58	12.16	12.74
Slaughter premium, EUR/animal		23.17	41.53	25.78	22.3	20.27

* average premium amounts; Source: LAEI data.

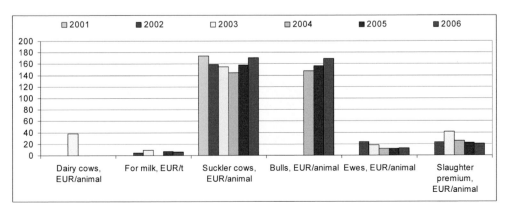

Figure 4. Dynamics in support level 2001-2006.

Effect of milk quota on farm management

It is not expected any important changes in management practice under quota system implementation.

As payments for suckles cows and bulls will be growing, it is expected that the number of meat breed animals will rise. More animals will be reared for meat production; however, it is not the effect of milk quota introduction.

There is no for a while free transfer of quota. The transfer (NOT SOLD) of quotas can be executed only in case the farm is inherited or the farm is sold. There is no for a while any quota exchange bureau in Lithuania. Quotas in the country are linked to animals. The transfer of quotas to a central agency can be executed only in case the farm stops its activity, i. e. falls out from milk production system.

As quotas in the country can not be sold, there are no quotas prices identified. There is no any instrument of leasing in Lithuania.

In 2004 there haven't been applied any reduction percentage – the quota agreed under Act of Accession haven't been distributed fully. There is a difference between the cost structure in large farms and small ones (see Figure 5).

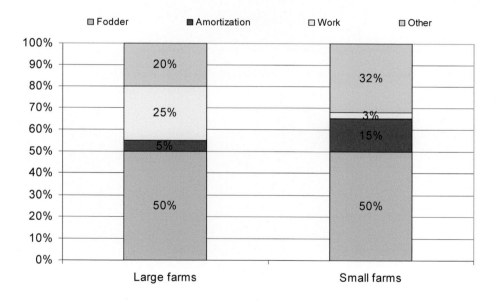

Figure 5. Cost structure in small and large farms.

The biggest challenge for the dairy industry in years ahead in Lithuania

From the year 2007 the transitional period for implementation of milk procurement will be over. Thus, all milk producers will be fully engaged to respect milk procurement requirements. According today's tendencies, it is expected that the important part of milk producers (the smallest ones) won't meet these requirements. The small producers have not sufficient investment potential in order to make needed investments in their production sale systems.

The forecast of milk production expenditures is provided on the basis of the following preconditions: growth of investments into modernization of the milk sector, intensification of livestock productivity and enlargement of farms increase of labour efficiency and income levels. Along with the modernization of the milk sector, labour productivity is expected to increase. With the aim of increasing milk production, better-balanced feeds will be required. Therefore, expenditure on feed will increase. Depreciation costs for equipment and buildings will undergo major increase due to the modernization of the sector.

Figure 6. Holstein-Friesian cows on the pasture moving to the West (EU).

Farm management and extension needs under the EU milk quota system in Slovenia

Stane Kavčič

University of Ljubljana, Biotechnical Faculty, Zootechnical Department, Groblje 3, 1230 Domžale, Slovenia

Introduction

With Slovenian accession to the European Union (EU) introduction of supply restrictions in milk production is becoming a fact. Milk quota, allocated to Slovenian farmers at national level was at the top of accession questions both for responsible government bodies and for farmers associations. Furthermore, distribution of milk quotas to individual milk producers was and still is a great fear of dairy farmers due to its direct economic consequences as well as due to its psychological impact related to high uncertainty of their likely future in dairy farming.

Data concerning the number of milk cows and milk production of individual dairy farms in Slovenia is not transparent. According to official data from agricultural census carried out in 2000 Slovenian dairy cows population counted 141.6 thousand distributed on more than 28 thousand farms (SORS, 2002). However, data for the same year provided by Slovenian dairies (Business association for Slovenian food industry) reported approx. 118 thousand dairy cows on 17 thousand farms included in milk delivering[8] (cit. Osterc *et al.*, 2001). Nearly half of those -mainly family farms - reared 4 dairy cows or less, with additional 30% of dairy farmers with herds of 5-9 cows. Economic or better to say income interesting milk production in Slovenia started in nineties at moderate to high milk yields (more than 5,000 kg per year) in a range of over 20 dairy cows (Lipovsek, 2002). This limit is expected to shift even much higher in the foreseen producer price reduction (Erjavec *et al.*, 2002). In these circumstances roughly 10% of farmers - those with herds of more than 15 dairy cows - have real prospects in milk production in the future. Additional 10-15% of currently milk delivering producers can also expect to achieve important share of their income from milk production also in the future. But farmers with less than 10 dairy cows also claim for as high milk quota as possible. Current Slovenian agricultural policy does not have any efficient program to speed up necessary structural change in direction of herd size improvement, nor idea about adequate system of accompanying measures to release expecting pressure for milk quotas by small farmers not really have any prospects in milk production.

In this short contribution some basic economic facts of Slovenian milk production are presented first. From this facts and economic projections for near future (mid-term outlook) new challenges for dairy farmers are foreseen. The difficult task of adjustment to new economic environment is challenge also for advisory and other supporting services in the whole dairy chain. This paper aims to present some opportunities for alleviation of milk quota burden. In this way it could contribute to the discussion of milk quota management, which would be best suited to Slovenian circumstances.

---

[8] In 2004, the first year of milk quotas in Slovenia, the number of milk delivery farmers decreased to approx. 11 thousand (*electronic database of Slovenian paying agency*).

Changing economic environment

According to CMO for milk and dairy products reform from 2002-03 there is clear trend of decreasing milk price in the near future. With introduction of milk premium that will increase in three steps till 2006 price cuts will be partially compensated. Anyhow, revenue per litre of milk produced is anticipated to fall, while opposite is expected with input costs (Table 1). In such an aggravating environment profitability of milk production can be sustainable only with increased milk yields and increased milk production per farm. But how to deal with this problem under quota regime imposed?

Table 1. Basic economic facts and projections for milk production.

	1992	1996	2000	2002	2004	2006	2008	2010
Cow's milk (3.7% fat) price €/100 kg	24.29	24.88	29.05	28.87	24.53	23.72	23.49	23.49
Compensation €/ha	11.59	0.00	9.87	14.81	22.22	33.32	263.63	263.63
Milk revenue €/100 kg	**24.42**	**24.88**	**29.16**	**29.04**	**27.15**	**26.64**	**26.58**	**26.58**
Dairy input cost index	0.98	1.04	1.05	1.06	1.07	1.09	1.12	1.15
Milk revenue/input cost index	**25.02**	**23.93**	**27.68**	**27.47**	**25.41**	**24.39**	**23.69**	**23.04**
Beef cattle revenue/input cost index	27.19	24.25	25.72	24.79	27.37	28.12	27.48	27.51

Source: own estimations (national agricultural sector model)

At national level the number of dairy cows will decrease further - from 210.000 in 1992 to 95.000 in 2010. Reduced number of dairy cows should be almost completely replaced by suckler cows. This seems realistic projection only in the case that suckler cows' premium will remain coupled also after CAP reform implementation[9].Along with this process average milk yield is anticipated to double. Total milk production increased over the last decade but seems that already reached it pick and that it will decrease down to the level of mid nineties (Table 2).

Table 2. National figures for Slovenian cattle sector.

	1992	1996	2000	2002	2004	2006	2008	2010
Milk production per cow	2,672	3,526	4,525	4,892	5,199	5,512	5,808	6,085
Dairy cows ending	211.81	166.12	137.72	127.94	119.01	107.22	100.63	95.86
Suckler cows ending	11.40	45.35	74.61	82.76	91.91	103.07	109.44	114.12
Total beef and veal production	47.13	51.33	49.20	47.39	48.58	49.38	49.26	49.17
Total milk production	566.05	585.76	623.23	625.87	618.74	591.03	584.51	583.29

Source: own estimations (national agricultural sector model)

---

[9] CAP reform implementation will take place in Slovenia in 2007. System of reform that will be implemented is not yet defined.

Farmers' responses to new circumstances

National averages are, however, of little importance for individual farmer. His farm income in the environment of limited production can be sustained only with increased production per unit of labour input or lower costs per unit of production. This can be possible with different farm management strategies e.g. taking over another farm, buying or leasing land with milk quotas or only milk quotas (after transition period). Under current production structure continuous increasing of milk output per farm is necessary to be more competitive. Other farm management practices are also possible including lowering forage costs and increasing efficiency of forage utilisation, lowering replacement costs e.t.c. Income diversification practices are another alternative on many farms with no prospects in dairying. This includes intensification of other enterprises on dairy farms (adapted crop rotation, calf rearing, beef and sheep, also income diversification with non-land-using and/or non-farm enterprises). Farmers should make use of different options. They are forced to dynamically adapt to changed economic environment. There is no unique solution for farms with different objectives and resources.

Farmers can make use of farm accounts as an aid to their management. Application of farm planning techniques is there a possible short cut to the progress. Farmers should rethink about missed opportunities including interesting options offered by agricultural policy measures. Careful implementation of farm plan, often in many small steps in technology of milk production, can lead to success. In his day-to-day practice farmers should think again and again about different options of reducing unit costs of production. From month to month or at least annually they should compare achieved results with budgets. By doing this, successful dairy farming is not a routine. It is a challenge for knowledgeable people.

Milk quota as new challenge for advisory service

Milk quotas are new challenge also for extension service. Although the system of milk quota in the first milk year after its introduction (2004/2005) did not function completely in accordance with CMO, farmers need clear advice in their unique circumstances. General instructions have very limited value for them. New system indispensably induced that only few advisors have enough knowledge to offer useful service to professional dairy farmers. In this new environment partner relationship would be desired, even necessary. Work in small groups for deepened work with individual dairy farmers seems good potential starting point. Above all farmers expect increased administrative support. This will become and remain 'new' task of extension service.

Figure 1. Education for farmers.

For the end one would say that also in EU new member states (NMS) specialised advisors with technology and economic insights in dairy farming as desired but still very distant objective. They should act as farm business consultant. A way to this dream is still very long. However, this is not myth, but seems imminent reality also in Slovenia.

Challenges for agricultural policy and conclusions

Number of dairy cows in Slovenia would fall according to the trends during last five years in dairy sector, accelerated by price change effects due to dairy market liberalisation. This would result in a fall of dairy cows for milk delivering of approx. 10 to 15 thousand (Kavcic, 2002). Milk quota will induce further reduction of national dairy herd for approx. 15 thousand cows. Reduction of dairy cows population for 20 to 25 thousand (approx. 20%) in a period of 3 to 5 years is a new moment with tremendous consequences. Taking into account investments in dairy farming during last years, which were in reference year or still are not yet completely utilised, pressure on undistributed national reserve of milk quota is expected to be remarkable. All these problems force dissatisfaction of farmers that are dependent on milk production and lead them to the question: "Is there any rational solution?"

The answer is not straightforward. To achieve satisfactory income in dairy farming, farmers in Slovenia in 2001 had to have at least 30 dairy cows with yields of 5,000 kg per year or more. This limit moved up very fast during last years. It is expected that similar economic tendency will be the fact also in the future (Lipovsek, 2002). Considering these figures there is room for only 3 thousand professional dairy farmers in Slovenia. As professional dairy farming is not prevailing feature of Slovenian milk production, it is expected that much more than economically justified number of farmers will sustain in milk production, in milk delivery and therefore also in milk quota distribution. Many of them, however, will be confronted with negative economic results.

Returning back to discussion about possibilities to mitigate income reduction dilemmas - foreseen as Slovenian EU accession effect, but seems to be unavoidable in policy environment of reduced policy intervention as long term trend - there is not much room for any speculations. Even when there is illusion that CAP accompanying measures can have important impact on dairy farmers' income situation, many of them are likely to be only dream of farmers in EU NMS (i.e. non-discriminative level of direct payments) or - if they will be eventually applied - they will be of transitional nature and there will be difficult to fulfil eligibility criteria (early retirement scheme). Income assistance measures that seem to have most sustainable nature are those from environment box of CAP, with some characteristics of animal welfare and food safety. Most of them are from technological point of view connected with sufficient agricultural land for manure disposal and therefore require limited number of animals per unit of land cultivated.

Crude estimates of likely EU accession effects for Slovenian milk production sector (Kavcic, 2002) clearly show that expected change of economic environment for dairy farmers in Slovenia is not promising. Although income per dairy cow could remain relatively stable, income in the sector will drop substantially due to very restricted milk quota imposed. Therefore, many of farmers with small dairy herd are expected to get out of production. To mitigate their decision for voluntary quota transfer to those with real prospects in dairy sector, compensation of income loss seems economically justified. Along with expected increase of budget expenditure to those remaining in production (LFA payments, environmental programmes) some additional measures would be of high importance. In addition to early retirement scheme, which is intended to farmers of relatively narrow age group, more targeted voluntary supply restraint scheme in a transition period of 5-10 years would be in Slovenian circumstances of remarkable value to faster necessary structural changes in direction of competitive milk production structure.

Figure 2. Expert trip about effects of agricultural policies on dairy sector in Slovenia.

References

Erjavec, E., T. Volk, S. Kavcic, M. Rednak & L. Juvancic, 2002. Ocena pogajalskih izhodišč Evropske unije na področju skupne kmetijske politike (Estimation of European Union negotiation position in the field of common agricultural policy). Expertise. Ljubljana, Agricultural Institute of Slovenia.

Kavcic, S., 2002. Introduction of milk quotas in Slovenia: Possibilities, accompanying measures and expected outcomes. Paper presented at International Conference 'Animal Science Days'.

Lipovsek, B., 2002. Gospodarnost prireje mleka v Sloveniji v obdobju 1995 - 2001 (Economic position of milk production in Slovenia during 1995-2001). Unpublished B.Sc. thesis. Domzale, Dept. of Animal Science.

Osterc, J., M. Klopcic & I. Valjavec, 2001. Strukturne spremembe v prireji in prodaji mleka v zadnjih dvajsetih letih (Structural changes in milk production and delivery in last 20 years). Sod. kmet., 34. 307-314

SORS (Statistical Office of the Republic of Slovenia), 2002. Agricultural Census 2002 data. Ljubljana, SORS.

Milk quota system in Latvia and impact on dairy sector

Inge Grinberga

Ministry of Agriculture, Republikas Iaukums 2, 1981 Riga, Latvia

Introduction

Dairy sector in Latvia is one of the most important sectors of agriculture and still represents about 45% of total final production of livestock products. The dairy sector has undergone the transition to the principles of a market economy during the last fourteen years and sharp changes occurred in milk production and in processing industry. Milk quota system by limited amount of production brings on further impact on structural changes in dairy sector in order to ensure the adequately high milk price.

For Latvia milk quota amount is set 695,395 tonnes till 2006 and from year 2007 of which 468,943 tonnes is delivery quota with a representative fat content 4.07% and 226,452 tonnes direct sales quota. Additional 33,000 tonnes is to be added from 2007. To the date, delivery quota is allocated to individual purchasers up to 98% of total amount but direct sales quota 44% of total amount. National quota reserve is set 1% for both delivery and direct sales quota. The milk production level for coming years till 2007 is estimated about 820,000 tons taking into account the allocated milk quota for Latvia (Figure 1).

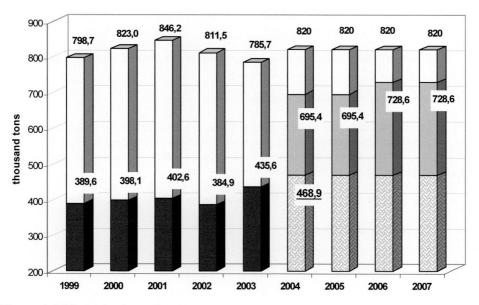

Figure 1. Milk production and quota.

The fixed amount of production ceiling will impact dairy sector – the number of dairy cows will keep the trend to decrease. Comparing to year 2000 number of cattle herds has decreased for 18% but number of dairy herds for 15%. The current dairy herd size is 5.8 cows per herd which is comparatively small size and it is foreseen that in coming years dairy herds will be increased. Nevertheless, it is envisaged that production figures will increase due to stable increase in productivity. The average productivity of cows under milk recording since 2000 has increased for 9% reaching 4,791 kg per cow in 2003 (Figure 2). Also positive tendency can be seen in the increase of cows under milk recording - comparing to year 2000 the number of milk-recorded cows has increased for 25%.

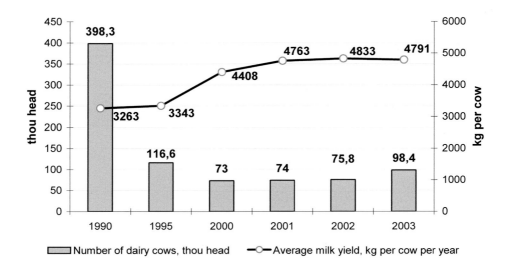

Figure 2. Dynamics of dairy cow numbers and average milk yields.

One of the unfavourable factors of milk production in Latvia still is the seasonality of milk production. Both the producers and the processing industry suffer from it since the excess supply of raw milk in the summer period milk prices hit the bottom levels whereas milk shortage is experienced during winter and autumn seasons.

Table 1 shows the relatively stable and positive trend of milk processing and the average price in Latvia in the period 2000-2003. It has increased in 2003 by 9% over 2000.

Table 1. Milk processing and average price of milk (2000-2003).

	2000	2001	2002	2003
Processed milk, 1000 tonnes	398.1	402.6	384.9	435.6
Average price, LVL / ton	87.20	95.52	94.09	96.09

Source: Central Statistical Bureau

Since the end of last year the milk price is growing and the price of first quarter of 2004 is 122,47 Ls/t.

Since 1996 Latvia has allocated certain state support in the form of direct and indirect subsidies to milk producers. It was mainly focused on breeding activities with the aim to increase total milk production in the country by modernization of farms and increasing productivity of cows. After the accession national support is allocated in compliance with the Accession Treaty and Community Guidelines for state aid in the agriculture sector Art.15. Table 2 shows that sheep sector is eligible for national support in compliance with Council Decision No. 281 of 22 of March 2004 in addition to the EU direct payments.

The EU direct payments and Complimentary national direct payments will have definite influence on structural changes in the dairy sector (Table 2). In particular, payments in beef sector give farmers a possibility to change farming from milk production to beef production in circumstances when production appears suitable for it.

Table 2. National support in animal husbandry sector.

Position	Before Accession, (EUR)	After accession				
		State aid[10] (EUR)				CNDP[11]
	2003	2004	2005	2006	2007	2004
Sheep sector	26,679	107,000	85,600	64,200	42,800	244,000
Goat sector	1,250	-	-	-	-	-
Dairy sector	5,760,698	-	-	-	-	4,048,000
Bovine sector		-	-	-	-	
Suckler cow pr.	75,491	-	-	-	-	2,684,000
Slaughter pr.	2,464,333	-	-	-	-	9,946,000
Fodder area	-	-	-	-	-	7,075,000

Milk quota system brings on two big challenges for dairy farmers – management and cooperation. Cooperation is essential not only between dairy farmers themselves but also between dairy farmers and processing industry, dairy farmers and breeding organizations. Taking into account the transitional period and the tremendous changes in the sector during the last fourteen years it is evident that the sector is still developing. First of all it was a matter of changing the thinking of farmers meanwhile learning new management practises and necessity to cooperate. Farmers had to learn to think in a strategic way, to plan and analyze costs and incomes, improve feeding technologies thus to benefit from better quality genetics. We can foresee that the productivity of dairy cows will be increased and breeding organizations will carry on the approved breeding programmes by introducing high quality genetics in herds. So far no breed replacement or any other significant changes in breeding programmes will be introduced on short term. Currently there are three breeding programmes in dairy sector end they cover about 200 thousands dairy cows out of which 75% are Latvian Brown breed cows and 24% Black and White Holstein breed and about 1% are other breeds like Danish Red, German Red and Latvian blue breed cows. The main breeding goal is to reach high productivity and strong body conformation cows, which has good health and providing economic milk producing.

The implementation and supervision of milk quota system in Latvia is done by Pedigree Information Data Processing Centre. There is also established a committee that consists of

---

[10] State aid – Council Decision of 22 of March 2004 No. 281
[11] CNDP – Complementary national direct payments

representatives from dairy farmers organizations, dairy processors organizations, Rural Support Service and data processing centre. The committee is responsible for dealing with issues related to milk quota allocation from national reserve and approval of milk purchasers. The farmers can apply for milk quota from national reserve if they have increased the herd and the application can be made three times during milk quota year.

Within frameworks of milk quota system farmers have the possibility to sell and buy quota with or without land from farmer to farmer and also there is a possibility for farmer to convert direct sales quota into delivery quota without any restrictions. In case if farmer wants to obtain milk quota, he must have milking cows registered in the Animal and Herd Register. For the moment there are few cases of buying and selling quota so it is not clear what average price for milk quota is.

There are few restrictions concerning milk quota leasing. The minimum amount for leasing of delivery quota is 1,000 kg but for direct sales quota 500 kg. Leasing is possible for two consequent years, afterwards the farmer have to decide whether to restart milk production, to sell quota or to transfer it to national reserve.

Data Processing Centre is responsible for milk quota system execution and implementation, that is, maintaining of milk producers register, collecting and processing information of delivery and direct sales quota, registration of all transfers and changes of milk quota. State Animal Breeding Inspection controls milk quota system – milk producers, milk purchasers, milk laboratories, and Data processing Centre.

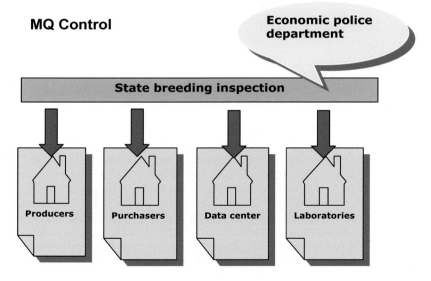

Figure 3. Milk quota control system.

The Latvian Agricultural Advisory and Training Centre is a non profit organization with 99% of principals owned by state and 1% Latvian Federation of Farmers. Advisory centre on the bases with contracts with the ministry of Agriculture provides farmers with information not only about milk quota system but also provides services in accounting and finances, information about economics and rural development and organizes practical training with participation of qualified foreign and Latvian staff. 228 advisers and specialists of centre provide consultations, services and trainings in plant production, animal husbandry, economics and other areas. Milk quota seminars were held in every region both by Advisory centre and by Data Processing Centre in order to reach as maximum amount of audience as possible.

Figure 4. Advisory offices in 26 districts of Latvia.

Milk production in Latvia still is rather fragmented since 1990 due to privatisation process and the abolition of large farms. In further ten years dairy sector in Latvia has to continue with structural changes in milk production sector and processing industry by centralization and concentration in milk production and processing in Latvia, which is contributing to minimizing production costs as well as to improving milk quality.

Farm management and extension needs under milk quota system in the prospect of Romania's entry into the EU

Agatha Popescu

University of Agricultural Sciences and Veterinary Medicine Bucharest, Faculty of Management, Economic Engineering and Rural Development, B-dul Marasti 59, sector 1, 011464 Bucharest, Romania

Number of cows, milk production, average milk yield and farm structures between 1990-2003

After 1989 the Romanian agriculture, including dairy farming has passed through a period of transition characterized by deep changes concerning ownership, production technologies, farm structures, technical endowment, labour force, productivity, agriculture contribution to GDP and agro food domestic and foreign trade.

The dairy sector has recorded a deep decline concerning cattle and cow stock, milk production, farm structures, degree of mechanization, applied technologies, productivity and profitableness, milk quality, milk delivery with an important impact on milk processing industry. The demand/offer ratio has become unbalanced in favour of demand. The opening of the borders has encouraged milk products imports.

The number of cattle has decreased by 54.26% from 6,291 thousand head in 1990 to 2,878 thousand head in 2003. The number of dairy cows and buffalos has also decreased by 28.73% from 2,468 thousand head in 1990 to 1,759 thousand head in the year 2003 (Figure 1).

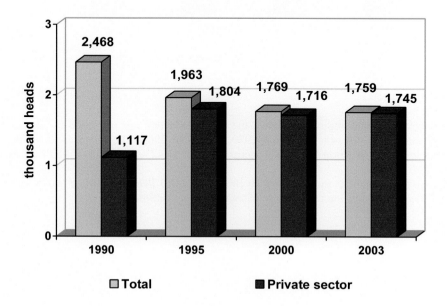

Figure 1. Number of cows, buffalos and heifers.

This was the result of the dissolution of the agricultural production co-operatives, numerous slaughters and severe culling percentage in order to select the highest productive animals. Another positive aspect is the increasing number of cows, buffalos and heifers within the private sector, so that in the year 2003, 99.20% of milking animals belongs to private farms and households in comparison with 45.25% in the year 1990.

The share of cows, buffalos and heifers in the total number of cattle within the private sector has decreased from 53.65% in 1990 to 61.29% in 2003.

Milk production has recorded a variable level along the analyzed period. However, an increasing trend of milk production has been noticed from 40,311 thou. hl in 1990 to 51,800 thou. hl in 2003 (Figure 2).

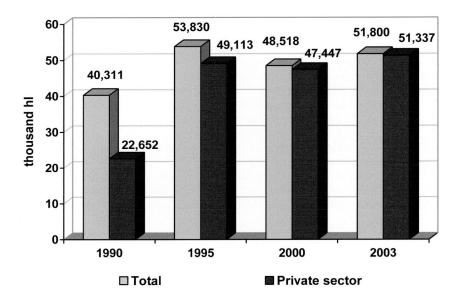

Figure 2. Milk production.

Considering calves consumption excluded, the milk production has increased by 36.07% from 33,057 thou hl in the year 1990 to 44,980 thou hl in 2003. The contribution of the private sector to milk production has deeply increased from 57.42% in 1990 to 99.10% in the year 2003.

The average milk yield has increased by 49.10% from 2,063 l/cow in 1990 to 3,076 l per cow in 2003, deeply influencing the increase of total milk production (Figure 3).

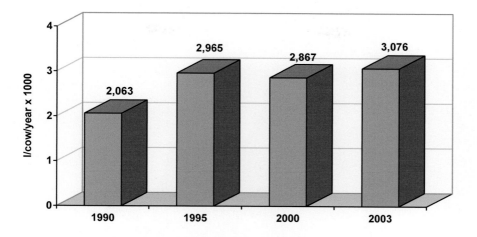

Figure 3. Average milk yield.

The number of cows and buffalos, milk production and average milk yield are different from a region to another (Table 1). In the year 2003, the contribution of various regions to milk production was the following one: North–Eastern Region 21.72%, North–Western Region 19.53%, South Region 15.83%, Central Region 13.68%, South-Eastern Region 9.83%, South-Western Region 9.83%, Western Region 8.48% and the remaining 1.10% belongs to Bucharest area. The highest milk yield is recorded in the farms situated in the surroundings of the capital and Ilfov county. An average milk production of 3,200 l/cow is performed in the farms placed in the North Western and Western Regions. In the Southern and South-Eastern areas, average milk yield is around 3,000 l/cow. In the North-Eastern and South-Western parts of the country, the average milk yield is less than 3,000 l/cow. There are significant differences between regions concerning the number of cows. The highest number of cows is raised in the North-Eastern area (406 thou. capita) and the lowest one, around 14 thou. capita in the proximity of the capital. There are also differences regarding average milk yield and milk production between counties belonging to the same region and between regions as well.

Table 1. Number of cows and buffalos, milk production and average milk yield by region in 2003.

Specification	M.U.	Total	Region							
			North East	South East	South	South West	West	North West	Centre	Bucharest
Cattle, of which:	Thou. heads	2,878	678	327	414	316	239	489	395	20
Dairy cows and buffalos	Thou. heads	1,759	406	181	268	185	142	324	239	14
Average milk yield	l/heads	3,076	2,949	3,048	3,077	2,816	3,227	3,273	3,084	4,326
Milk production	Thou. hl	51,800	11,252	5,094	8,202	5,092	4,394	10,118	7,088	560

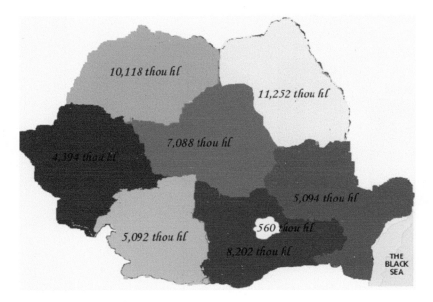

10,118 thou hl

11,252 thou hl

7,088 thou hl

4,394 thou hl

5,094 thou hl

560 thou hl

5,092 thou hl

8,202 thou hl

THE
BLACK
SEA

Figure 4. Regional differences in the development of dairy sector.

The North East, North West, Centre, South and Bucharest regions will continue to be the most important milk suppliers. The hilly and mountain areas will produce the cheapest milk and this milk will be mainly biological milk.

In the year 2003, there were 1,189,124 dairy farms, by 13% less than in 1995. The farm structure is presenting as follows: 93.58% farms are growing 1-2 cows, 5.63% farms have between 3 and 5 cows and the remaining 0.79% have more than 5 cows. Only 176 dairy farms are raising over 100 cows (Table 2).

Table 2. Dairy farms structure.

Farm size group	1995		1999		2003	
Heads	Farms	%	Farms	%	Farms	%
1-2	1,318,384	96.54	1,169,552	95.00	1,112,839	93.58
3-5	43,089	3.15	55,736	4.52	66,959	5.63
6-10	2,537	0.18	3,899	0.32	5,833	0.49
11-15	419	0.03	787	0.06	1,542	0.13
16-20	202	0.01	304	0.02	824	0.07
21-30	161	0.01	223	0.02	442	0.04
31-50	142	0.01	168	0.01	309	0.03
51-100	116	0.01	117	0.01	200	0.02
Over 100	491	0.03	295	0.03	176	0.01
Total	1,365,541	100.00	1,231,081	100.00	1,189,124	100.00

Source: Ministry of Agriculture, Forestry and Rural Development

As a positive aspect of the continuous structural changes, a trend of reducing the number of dairy farms can be noticed, in general, at the same time with a slowly increase in average herd size. While the number of the smallest farms raising 1-2 cows is decreasing, the number of the largest farms growing over 100 cows is also diminishing. In the year 2003, there are 4.07 times more farms raising 16-20 cows, 3.68 times more farms growing between 11 and 15 cows, 3 times more farms with 6-10 cows, 2.74 times more farms with 21-30 cows, 2.17 times more farms raising 31-50 cows and 1.72 times more farms with 51 up to 100 cows.

The number of dairy cows has decreased by 11.35% from 1,964 thou. heads in 1995 to 1,741 thou. heads in 2003. In 2003, 77.70% of total cows number belongs to the farms with 1-2 head, 13.66% cows are raised in farms with 3-5 capita, 2.40% cows are in the farms with 6-10 capita and just 2.18% cows are raised in the farms whose size is higher than 100 capita. In comparison with the situation in the year 1995, in 2003, an increased number of cows were noticed mainly in the farms raising 16-20 capita, 11-15 and 6-10 heads.

In 2003, the average number of cows per farm was 1.46 heads, as 93.58% of farms are raising 77.70% of the total number of cows, meaning that for this category, the average farm size is the lowest one: 1.21 cows/household (Table 3).

The main breeds grown in Romania are the following ones: 36% Romanian Yellow and White Spotted Bred (Simmental origin), 35% Romanian Black and White Spotted Breed (Friesian origin), 26% Brown Breed (Brown Swiss origin) and 3% Pinzgau Breed (Figure 5).

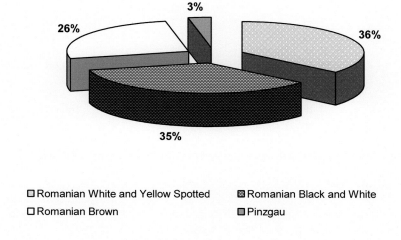

Figure 5. Breed structure in 2003.

Figure 6. Romanian white and yellow spotted cattle and Romanian brown cattle.

Table 3. Farm size in year 2003.

Farm size group Heads	Average number of dairy cows heads/farm	Share in total number of farms %	Share in total number of cows %
1-2	1.21	93.58	77.70
3-5	3.55	5.63	13.60
6-10	7.16	0.49	2.40
11-15	12.85	0.13	1.14
16-20	17.48	0.07	0.83
21-30	24.88	0.04	0.63
31-50	38.49	0.03	0.68
51-100	67.11	0.02	0.77
Over 100	216.36	0.01	2.18
Total	1.46	100.00	100.00

Structural changes in the near future – prospects for the period 2004-2010

Structure of dairy sector

Taking into account the evolution of the number of dairy farms, there are expectations as in the coming years the number of farms to be less numerous. Therefore, the prospect for the year 2010 is 1,069,206 dairy farms, by 10.09% less than in 2003. The figures shown in Table 4 were calculated based on the performances in 2003 and the annual decrease/growth rate recorded during the period 1990-2003 for each farm size group. There are expectations as the number of the smallest farms raising 1-2 cows to decrease by 13.09% and also the number of farms growing over 100 cows to decrease by 23.87%. At the same time, an increased number of farms raising between 5 and 100 cows is expected in the future, but mainly for the farms raising 16-20 cows (66% growth rate), 11-15 cows (63% growth rate) and 21-30 cows (55% growth rate).

Table 4. Prospects for the number of dairy farms.

| Farm size Group heads | Achievements 2003 | | Prospects | | | | | |
| | | | 2004 | | 2007 | | 2010 | |
	Farms	%	Farms	%	Farms	%	Farms	%
1-2	1,112,839	93.58	1,091,139	93.17	1,027,307	91.86	967,210	90.46
3-5	66,959	5.63	69,943	5.97	78,895	7.05	87,847	8.22
6-10	5,833	0.49	6,245	0.53	7,481	0.66	8,717	0.82
11-15	1,542	0.13	1,682	0.14	2,102	0.19	2,522	0.24
16-20	824	0.07	902	0.08	1,136	0.10	1,370	0.13
21-30	442	0.04	477	0.05	582	0.06	687	0.06
31-50	309	0.03	329	0.03	389	0.04	449	0.04
51-100	200	0.02	210	0.02	240	0.02	270	0.02
over 100	176	0.01	162	0.01	148	0.01	134	0.01
Total	1,189,124	100.0	1,171,089	100.0	1,118,280	100.0	1,069,206	100.0

Despite that at present, the number of dairy cows has relatively remained at a constant level for the first time during the last 14 years, it is expected as in the near future to record a new decrease so that in the year 2010 to reach 1,646,230 heads, by 5.48% less than in the year 2003. There are expectations as the number of cows to decrease mainly within the smallest farms raising 1-2 cows, but also in the largest ones growing over 100 heads. In all the other farm size groups, an increased number of cows is expected (Table 5). The figures shown in Table 5 were calculated based on records in 2003 and taking into account the growth/decrease rate registered by each farm size group during the period 1990-2003. An important increase (over 50%) is expected for the number of dairy cows within the following herd size groups: 16-20 heads (66.31%), 11-15 heads (64.77%), 21-30 heads (58.76%), 6-10 heads (50.81%).

Table 5. Prospects for the number of dairy cows.

| Farm Size Group Heads | Achievements 2003 | | Prospects | | | | | |
| | | | 2004 | | 2007 | | 2010 | |
	Cows	%	Cows	%	Cows	%	Cows	%
1-2	1,353,234	77.70	1,321,300	76.48	1,225,615	72.64	1,129,930	68.63
3-5	237,958	13.66	248,300	14.37	279,115	16.54	309,930	18.82
6-10	41,752	2.40	44,800	2.59	53,880	3.19	62,970	3.83
11-15	19,815	1.14	21,700	1.25	27,170	1.61	32,650	1.98
16-20	14,400	0.83	15,800	0.91	19,900	1.19	23,950	1.45
21-30	10,966	0.63	11,900	0.69	14,600	0.87	17,300	1.05
31-50	11,894	0.68	12,700	0.73	15,100	0.90	17,500	1.06
51-100	13,422	0.77	14,100	0.82	15,950	0.95	17,800	1.08
over 100	38,080	2.18	36,400	2.10	35,800	2.11	34,200	2.10
Total	1,741,551	100.0	1,727,000	100.0	1,687,130	100.0	1,646,230	100.0

The prospects for the average herd size structure are shown in Table 6 where the data could help us to draw the conclusion that in the year 2010 the average farm size could be 1.54 cows/farm, meaning a very slow increase by 0.08%.

Table 6. Prospects for the average farm size.

Farm Size Group Heads	Achievements 2003	Prospects		
		2004	2007	2010
1-2	1.21	1.21	1.19	1.17
3-5	3.55	3.55	3.54	3.53
6-10	7.16	7.17	7.20	7.22
11-15	12.85	12.90	12.93	12.95
16-20	17.48	17.52	17.52	17.48
21-30	24.88	24.95	25.08	25.18
31-50	38.49	38.60	38.82	38.98
51-100	67.11	67.14	66.46	65.93
Over 100	216.36	224.69	241.89	255.22
Total	1.46	1.48	1.51	1.54

Based on the average milk yield performed in 2003 and the annual growth rate recorded between 1990 and 2003, it is expected as in 2010 the level of this indicator to reach 4,242 l/cow/lactation, that is by 37.90% more than in 2003. Taking into account the prospects for the number of cows and average milk yield, in the year 2010, milk production will achieve 69,823 thou hl, a level by 34.79% higher than in 2003 (Table 7). It is difficult to make prospects by each farm size group, because at present, milk records are referring to just 8-10% of the total cow stock. However, it is known that in the smallest farms with 1-2 cows, average milk production is the lowest one in general, while in the larger farms have a higher level. But this is not a rule, because there are also farms with 6-10 cows or less with 4,500-5,000 milk l/cow, even more, depending on the farmer's managerial skills and mainly of his financial resources to raise the best cows.

Table 7. Average milk yield and milk production.

Specification	M.U.	Achievements 2003	Prospects		
			2004	2007	2010
Number of cows	in 1000	1,741	1,727	1,687	1,646
Average Milk Yield	l/cow	3,076	3,411	3,764	4,242
Milk Production	1000 hl	51,800	58,908	63,499	69,823

Concerning the structure of dairy farms according to the income coming from cows , it is difficult to give the right figures as long as, most of the farms/households have no farm records, book keeping and according to the legislation in force they are not obliged to set up balance sheet and profit and loss account. Just a few number of farms have a legal status and have farm evidence. Therefore, the figures presented in Table 8 are more or less approximated and have been established based on the reason that the farms raising more than 10 cows can assure a higher percentage of their income from marketed milk. As long as in Romania, the most numerous farms have less than 10 cows and are usually mixed farms raising also pigs, poultry, horses, sheep and cultivating agricultural crops, one could affirm that 99.50 - 99.70% of the total number of farms get less than 60% of incomes from cows.

Table 8. Prospects for the farm structure by income.

Income Level	Achievements		Prospects			
	2004		2007		2010	
	Farms	%	Farms	%	Farms	%
Less than 60% of income from cows	1,167,596	99.70	1,114,518	99.65	1,063,774	99.49
More than 60% of income from cows	3,493	0.30	3,762	0.35	5,432	0.51
Total	1,171,089	100.0	1,118,280	100.0	1,069,206	100.0

As we know, the higher the number of cows, the more activities in the farms, the higher costs. As 99.70% farms raise less than 10 cows, it is obviously that they are "family farms". In a farm with more than 10 cows one labourer is employed, taking into account that mechanization is missing, except milking. In a farm with more than 20 cows it is normal as 2 labourers to be employed. At present, there is no evidence concerning this aspect in dairy farming. Only 0.01% of the total number of farms, mainly raising over than 100 cows have employees.

Concerning milk production destination, one can say that in the smallest farms raising 1-2 cows, milk is used mainly for family and calves consumption and very rarely and a few amount of milk is delivered directly in the market. In larger farms, the higher the farm size, the more marketed milk. In average, around 15 - 20% of milk production is retained to cover family and calves needs and the remaining 80 - 85% is delivered to milk processing industry and directly to the market.

At present, according to the data provided by National Agency for Veterinary Services and Food Safety, there are 588 milk processing factories but just 19 are corresponding to the EU standards. A number of other 54 milk processing factories are preparing to cover the requirements imposed by Romania's entry into the EU in January 2007. Other 28 factories are going to pass thorough a 3 years period of transition from January 2007 on. The remaining milk processing plants have to meet the hygiene requirements, otherwise they could not continue to operate and produce for national consumption. The actual milk processing capacity is 3 million tons per year. Most of milk processing plants need investments in order to be modernized and apply the highest performance technologies assuring the obtaining of high quality fresh milk and other milk products according to the EU standards.

Regional differences in the development of Dairy Sector

Taking into account the evolution of the number of cows, dairy farms, average milk yield and milk production during the last 14 years and the actual situation in the country, there are expectations that the regional differences in the development of dairy farming to continue to exist. The North-Eastern, North Western, Southern and Central Regions will continue to supply the most important amount of milk. The regions with the highest average milk yield will continue to strengthen their milk producing capacity - Ilfov county and Bucharest area, North Western, Western, Central and Southern Regions.

Milk processing industry has to develop its processing capacity to keep pace with the prospects concerning milk production. New investments are required in order to implement HACCP quality system and other EU standards or getting competitive dairy products both for the internal and external markets.

Mixed farming or specialized dairy farms?

At present, a few number of dairy farms is represented by specialized farms. We are expecting as mixed farming, meaning dairy and crop production or dairy and horticulture or other sorts of combinations to persist for a long period of time in the Romanian agriculture, as long as milk price will be very low. Dairy farms will be in danger to fail if losses from high production costs will be not balanced by profit performed in other agricultural sectors.

Concerning part-time farming in dairy sector, we can talk about this aspect yet, due to the low grade of mechanization. Almost all activities within the majority of farms are manually done (feeding, cleaning, manure collecting, watering etc.). In a few farms, mainly with more than 20 cows, milking and watering are mechanized operations. In the farms with more than 100 cows, all the operations are mechanized assuring a high productivity.

Intensification or extensification of dairy farming?

During the last 14 years, the number of cows, buffalos and heifers per ha has continuously decreased as follows: from 17.3 heads/ha in 1990, to 13.5 heads in 1995, 12.4 heads in 2000 and 12.3 heads in 2003, meaning that in the last year the level of this indicator was lower by 29% than in 1990. In the year 2010 it is expected as the number of cows, buffalos and heifers per ha to be around 12 heads.

In the coming years, we are expecting an extensification of dairy farming, meaning less cows per ha in use, as long as the number of cows is expected to decrease. On the other hand, we are expecting a higher productivity in dairy farming, meaning that feedstuff produced on the same surface to better meet cows feeding requirements both from a quantitative and qualitative point of view and to assure higher milk performance.

Average milk price

Average milk price has slowly increased year by year from 0.11 Euro per litre in the year 2000 to 0.14 Euro per litre in the year 2004 (Table 9).

Table 9. Average milk price.

	M.U.	2000	2001	2002	2003	2004
Milk Price	Lei/l	3,800	4,300	4,500	5,200	6,000
	Euro/l	0.11	0.11	0.12	0.13	0.14

There are differences from a region to another and from a county to another, depending on a lot of factors of which the most important ones are:
• milk production level,
• milk demand/offer ratio,

- milk quality,
- milk processors.

Milk price is in close relationship with milk quality. This is the main reason that in the most of cases milk price is lower than production cost.

There are expectations that the milk price may increase in the coming years assuming milk quality meets the EU standards (Table 10).

Taking into account that milk cost is partially subsidized by the Romanian government (1,400 lei/kg, i.e. 0.034 Euro/kg), milk processors are tempted to offer a lower milk price.

Table 10. Prospects for Milk Price.

	M.U.	2005	2006	2007	2008	2009	2010
Milk Price	Lei/l	6,440	6,880	7,320	7,760	8,200	8,640
	Euro/l	0.15	0.16	0.17	0.18	0.18	0.18

The milk price will depend on: the milk demand, milk quality, season, regions and on the milk processors.

Farm management

1. Main problems in farm management

The most important aspect which must be taken into account is the fact that Romania, in comparison with other European countries, is facing a special situation:
- more than 1 million very small farms, which are not "real farms",
- in fact they are "peasant households",
- raising 1-2 cows , 1-2 pigs , hens, sheep, 1-2 horses and with a small agricultural land,
- the average farm size in the country being 2.5 ha, where usually peasants like to cultivate wheat, maize, barley and other crops , including forage crops.

Most of the actual farms are "subsistence farms" destined to cover family needs. Their main characteristics are the following ones:
- low technical endowment, limited to usual agricultural tools; no tractors, no installations, no irrigation systems etc in the most of farms;
- small agricultural land surface and divided into small plots, which can not allow to apply high performance technologies;
- agricultural works in the field and activities in dairy farming are manually done;
- low productivity both in dairy farming and crop production;
- lack of financial resources to assure farm inputs (equipment, fertilizers, seeds, high value animals);
- most of peasants are old people and the young generation is looking for jobs in the cities;
- farmers' low training level;
- reproduction is assured naturally, in a few farms frozen semen is used;
- cows are not controlled concerning milk performance;
- no book keeping, farm evidence.

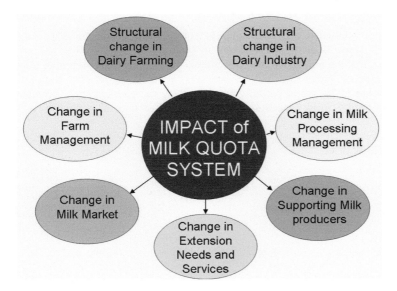

Figure 7. Impact of quota system.

There are expectations as the number of this type of farms to decrease in the coming years, but this is a long term process. At the same time this type of farms will remain in many parts of the country for covering family needs. The additional amount of milk, exceeding family needs will be not allowed to be sold in the market, as milk quality do not meet the EU standards. This sort of farmers have to join their animals, land, endowment in order to benefit of funding programmes to develop business in dairy farming if they would like to and to respect regulations concerning milk quota and milk quality.

There are also farms with more cows where we can meet a better situation as follows:
- high potential animals;
- more agricultural land – 10 ha and more according to the legislation in force; farmers received the ownership titles for their land and also use to rent land for producing a corresponding foodstuffs amount for covering cows feeding requirements;
- a better technical endowment represented by tractors, agricultural machines, milking machines etc;
- a higher training level and a continuous wish to get more knowledge and skills in dairy farming;
- cow reproduction assured both by means of AI services (average AI rate is 36% in the country at present) or naturally with tested local bulls;
- genetic gain is assured in cow population using frozen semen and high breeding value bulls;
- a few cows are in Milk Control, the average percentage is 8-10% at country level;
- foodstuffs are assured from the own land: green grass, silage, hay, roots, straw etc and for balancing rations concerning protein and energy content, farmers use to buy combined food, bran, sunflower cakes, residues from brewery industry;
- the farms situated in the hilly and mountain areas have pastures and meadows assuring a lower milk production cost, a higher quality milk production;

- the culled cows are replaced mainly with heifers born in the own farm, but also purchased from outside;
- a higher grade of mechanization (milking, watering, feeding, manure collection etc);
- a big problem is milk storage, most of farms have no milk tanks and possibilities to assure corresponding milk storage conditions;
- there are a few milk collecting centers in the local areas; usually farmers have to transport their milk to long distances to deliver it milk processors; a few milk processors use to collect milk directly from farmers;
- milk quality is also a problem as long as milk hygiene rules which are only partially respected; according to legislation in force, milk is collected based on the following reception criteria: fat %, protein %, acidity and not taking into account the number of pathogen germs and somatic cells; at present, more and more milk processors pay more attention to the last aspects, which are compulsory criteria for milk collecting in the EU countries;
- after retaining family and calves consumption (15 - 20%) of total milk production, farmers either deliver milk to milk processors or directly to the free market; all the farmers are looking for serious and solvable clients, able to pay the delivered milk as soon as possible and also to offer a higher price;
- incomes come in order from: marketed milk, culled cow and calf sold at 100 - 120 kg live weight, but also from other activities;
- incomes are able to cover production cost and to assure profit only in the larger farms raising more than 6 - 8 cows; for all the other small farms, incomes are not enough to cover costs or the profit level is very low, not allowing to develop business in dairy farming.

As a conclusion, the main problems of farm management are the following ones:
- lack of financial resources to assure farm inputs at a corresponding production level: high breeding value animals, high value seeds, tractors, machines, equipment;
- low technical endowment, except a few farms;
- farmers low training level; farmers need more knowledge and skills in the following fields: feeding, rational grazing, ration optimization, milk hygiene, reproduction, breeding, communication, EU quality standards, EU agricultural policies, programmes offering financial support like SAPARD, project management, business management, farm evidence and book keeping, environment protection; they also need to be better informed about legislation (veterinary regulations, financing, organization of associative farms).

2. Restraining factors in dairy farming:
- climate conditions: severe drought or rich rainfalls and floods affecting forage crops and obviously cow feeding during the last years;
- high prices for farm inputs (machinery, combined food, medicines, seeds, energy);
- a non stimulating credit system with high interest rates and a hard guarantee system; farmers are complaining of this aspect very much, they would prefer special credits for agriculture with 6 - 8% interest rate;
- milk processors are tempted to offer a lower milk price and farmers are obliged to accept it;
- besides income, Government subsidies are not enough to cover costs;
- non corresponding farm structure, too many small subsistence farms;
- a hard tax system for commercial companies.

3. Opportunities:
- a favourable demand/offer ratio, as it is not enough milk in the market and for milk processing industry;
- subsidies from the Romanian Government as follows:
  - for milk production,
  - for buying high breeding value heifers and cows,
  - for calves obtained from artificial inseminations,
  - for using selected and certified seeds,
  - for fuel purchase
  - for agricultural works,
  - for assurance premiums,
  - for buying agricultural machines, as compensation for damages;
- existence of Cattle Breeders Association at national and local level, involving in:
  - assuring farm inputs at lower prices,
  - farmers' training,
  - technical assistance,
  - milk price negotiation with milk processors,
  - looking for clients,
  - milk collecting points,
  - diversified milk records,
  - farm evidence,
  - organization of animal fairs and exhibitions,
  - representatives of farmers' interests ;
- in the hilly and mountain areas one can discuss about "biological dairy farming", which could be an alternative for a higher milk price;
- existence of local traditions in milk products which could be used in rural tourism, an important additional income source for dairy farmers in the hilly and mountain areas.

According to the last events concerning Romania's entry into the EU, Romania proposed to the EU a 5 billion tons milk quota, taking into account the amount of marketable milk, but it was accepted just a milk quota of 3,057 billion tons, level which can be easy covered based on the records from the previous years. The distribution of this milk quota will be: 1,093 billion liters for milk processing industry, 1,964 billion liters will be sold directly on the free market and 188,400 million liters – a reserve for the year 2009.

The main aspects farmers have to pay attention to under the milk quota system are the following ones :
- to produce as much milk as quota allows, because for the additional milk farmers will have to pay a fine;
- farmers who are not able to cover the allotted quota will have to sell their quota or to rent quota; therefore milk quota market will have to appear and farmers have to be aware of this;
- this means to assure the best correlation between the number of cows, their genetic production potential and feeding, as long as feeding is the key factor for getting a high milk performance;
- a corresponding feeding according to cows feeding needs, the number of cows, their live weight, production level, pregnancy period, etc.;
- a better correlation between land surface, crop structure and production from a qualitative and quantitative point of view and rations optimization concerning protein, energy and cost level;

- reproduction based on AI services, corresponding matings with high breeding value bulls, cow assistance during pregnancy period and care for parturition and suckling calf;
- cow breeding based on special breeding programmes based on milk records, severe cow selection, use of the high breeding value bulls, bull testing;
- milk quality, depending on foodstuff quality, milk hygiene, cow and shed hygiene;
- adopting the high performance technologies assuring a high grade of mechanization, milk quality and productivity;
- investments based on EU and Romanian Government funds for building new modern farms of an optimum size (20 cows) with a corresponding technical endowment, high performance animals, high productivity, efficiency and milk quality;
- firm milk contracts concluded with milk processors;
- the moment when Romania will entry the EU structures (January 2007), the Romanian farmers will benefit of direct payments for 150,000 suckling cows (premium) which are not able for supplying milk or milk products for a 12 months period. The premium value is 200 Euro per cow.

Romania's entry into the EU will result in the following developments and changes:
- change of dairy farm structure in the sense that the smallest farms will disappear and the most appropriate herd size assuring profitability will be around 20 cows;
- change of farms in the sense of being "more market oriented", becoming "commercial farms";
- new investments will be made in dairy farming, mainly in the hilly and mountain areas, close to cheap forage resources;
- new high performance technologies will be applied for assuring high productivity and high milk quality;
- improvement of productivity in pastures and meadows, rationale grazing on pastures and a better maintenance of meadows in order to produce cheaper milk;
- biological dairy farming will be developed in the hilly and mountain areas; the first biological dairy farm in the country called La Dorna in the Northern Romania was founded in the year 2004; other 151 modern dairy farms will be founded in the coming years, buildings, equipment and heifers being 50% from the Romanian Government;
- financial support for farm modernization, restructuration and farmers' training (SAPARD, FIDA, etc);
- extend of Milk Control and AI system for assuring milk evidence and genetic gain;
- special financial support for young farmers in order to set up modern dairy farms;
- improvement of agricultural extension services delivered by National Agency for Agricultural Extension and its network at local level, agricultural universities, research institutes and stations, SAPARD and World Bank Programmes;
- modern and high performance endowments for Milk Control Laboratories, mainly concerning the number of pathogen germs and somatic cells;
- advantaging credits with a low interest rate for dairy farmers;
- premiums, allotments for buying farm inputs: high value animals, machinery, etc;
- preservation of local traditions concerning specific Romanian dairy products; a number of 58 types of Romanian cheese and other dairy products made of cow, sheep, goat and buffalo milk, obtained by traditional technologies were approved by the EU Commission;
- a more efficient milk collection, transportation and storage in order to assure high quality of milk as raw material for milk processors;
- modernization of milk processing factories in order to meet the EU standards (implementation of quality management system);

- assurance of animal welfare;
- environment protection against pollution using the modern technologies "friendly" with environment ;
- more care for food safety, producing healthier products;
- more involvement of Breeders Association in helping farmers to join, to assure farm inputs, to deliver milk;
- support for young farmers: government funding, low interest credits, certified training.

New management practices

The implementation of milk quota system in Romania will create serious problems to farmers and will require to get more managerial knowledge and skills in the following directions:
- milk quota system;
- business planning;
- budgeting;
- milk marketing;
- book-keeping and interpretation of balance sheet and profit and loss account;
- implementation of integrated information systems at farm level: computer skills and IT facilities assisting the farmers managerial needs in creating data base, data processing (costs, gross margin etc), decision making, optimizing resource allotment etc;
- communication and negotiation skills;
- time management, solving problems skills and a better work organization;
- new production technologies.

Cost structure

Milk production cost is a key indicator reflecting efficiency in a farm. The most important part of cost structure is represented by variable costs as follows: feeding cost 64 - 70%, replacing heifer 8 - 8.9%, veterinary services 1.4 - 3.5%, A.I. cost 0.6%, watering, electricity, heating 1.8 - 2.6%, subscription to Local Breeders Association and taxes 0.2 - 1.2% , labour 8.5 – 16%, depending on farm size, other costs 4.6 – 6%. These figures can be more or less considered as an orientation, because from one farm to another costs are different depending on local conditions: number of dairy cows, production level, technical endowment, labour, but also on farmers 'managerial skills how to keep costs under control. It is obvious that the largest cost factors are in order: feeding, replacing heifer, veterinary services. Fixed costs are important only in the largest farms.

Organization of extension and Extension needs

National Agency for Agricultural Extension and its local network

The National Agency for Agricultural extension, called ANCA, was founded in 1999, according to the ADAS International British and ANDA French Models, adapted to the conditions in Romania and based on a PHARE project with technical assistance assured by the EU experts. The body is linked to the Ministry of Agriculture, Forestry and Rural Development (MAFRD) and has 41 subsidiaries at county level, within the 41 Agricultural Directorates of MAFRD. Each subsidiary has several agricultural chambers, one chamber in every 3 communes, in order to enable extension officers to deliver consultancy and technical assistance to farmers living in the villages belonging to those communes.

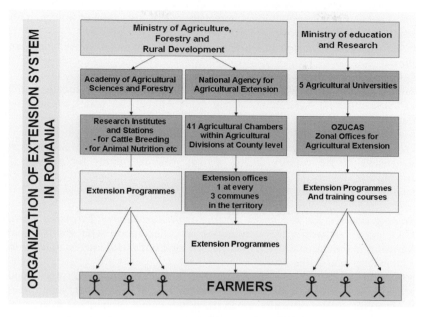

Figure 8. Organization of Extension system in Romania.

At present an extension officer assures extension services for 5 - 6 communes, but for the year 2007 it is planned only one extension officer, meaning an increasing number of qualified consultants.

ANCA network is running its activity according to the programmes of agricultural development and agricultural policies established by MAFRD and is financially supported by the Romanian Government and EU funds.

The extension staff has a high qualification, graduates of the Agricultural Universities with a long practice in agriculture, high competence both in technical and economical aspects (farm management, agro-food marketing, environment protection, sociology. In the year 2002, a number of 300 extension officers was trained abroad. The 5 Agricultural Universities existing in the country offer special Master Courses (2 years length) for qualifying the young graduates but also to help the actual officers to update their knowledge and skills. A special training programme for the Romanian extension staff is assured by the EU experts from the University of Hohenheim, Germany. At university level, regional offices for Agricultural extension have been created in order to assure periodical extension staff training.

A.N.C.A. has a close collaboration not only with the agricultural universities, but also with the research institutes and stations subordinated to Academy of Agricultural Sciences and Forestry in order to assure a corresponding and fast transfer of updated technological information and research results to farming practice.

Agricultural Universities involvement in Extension services

The 4 universities of agricultural profile in Bucharest (Cluj-Napoca), Iasi si Timisoara and two faculties of agricultural profile within the University of Craiova assure technical and economical assistance, extension services to all the categories of beneficiaries from agriculture and food industry. Every agricultural university has its own zone of influence for agricultural extension, but this is not a rule.

The activities of agricultural extension offered by the agricultural universities are diversified as follows:

A. *Training in Agricultural Extension for young specialists:*
- Agricultural Extension", daily and compulsory or optional courses according to the curricula of the faculties of: Agriculture, Management, Animal Husbandry, attended by full time students (5 years);
- „Master Courses" on Agricultural Extension for postgraduates, 2 years length.

B. *Programmes for Trainers' Training,* run in collaboration with various bodies involved in agricultural extension in the country or/and abroad, offering grants for advisors / consultants training

C. *Short-term training courses for improving students and farmers training level* in the EU countries;

D. *Short-term training courses for specialists dealing with Agricultural Extension Local Centres* (CJCA), organised within the universities in collaboration with the National Agency for Agricultural Extension (ANCA), on various topics of actual interest in order to improve their knowledge and skills.

E. *Extension services for all the farmers,* no matter their juridical status, type, profile and farm size, ownership sector; almost all these services are free of charge, but, some times they are paid according to the difficulty grade of technical and economical assistance. These services are the following types:
- at the farmers' request, depending on their concrete problems, raising within the agricultural practice and requiring optimum solutions to be urgently solved;
- within special programmes for agricultural extension destined to the farmers dealing with various agricultural branches (PHARE, B.M. MAFWE Projects).

F. *Exchange of knowledge, practices, ideas concerning agricultural extension,* on the occasion of the scientific symposia organized by UASVMs, where are invited to take part representatives from DADR, CJCA, Local Councils, managers of various companies, mainly from the counties placed in the university zone of influence;

G. *Participation of academic staff* to various meetings organised by DADR in collaboration with Producers Associations, CJCA such as: fairs, exhibitions of products and agricultural machinery, INDAGRA, a good opportunity to make demonstrations and exchange ideas;

H. *Dissemination of academic staff's research results* to specialists, students, doctoral candidates, researchers, farmers by editing and publishing university textbooks, brochures, practical guide lines, scientific papers, which are usually published in the University Bulletins, Newsletters but also in various agricultural reviews such as: Romania's Agriculture, Agricultural Profit, the Farmer, Cattle Breeder Review, Romanian Agriculturist etc.

I. *Special consultancy for processors* dealing with various Food industry branches

J. *Distance Learning* for candidates working in agriculture, but also in other fields of activity or are not employed and who would like to get qualification attested by an university diploma. We have to mention that during the last years, about 30% of Distance Learning students are farmers.

As an example you may find below concrete aspects concerning agricultural extension services delivered by *University of Agricultural Sciences and Veterinary Medicine Bucharest:*

A. *Forming specialists in the field of Agricultural Extension by*:
- *University training courses on Agricultural Extension,* organised for the students belonging to the Faculty of Management, Economic Engineering and Rural Development and the Faculty of Animal Husbandry. In the curricula of the 4-th year of study, this subject is compulsory or optional, providing 2 classes course and 2 classes practical (seminar). Until now a number of 397 graduates attended and promoted this subject.
- *Master Course on Agricultural Extension,* 2 years length, for the graduates of the Faculty of Agriculture, Management, Animal Husbandry and Horticulture. Until now, 2 series of young specialists counting 50 persons graduated this Master course and are working with ANCA and its local CJCAs.
- *Short term training courses, in Denmark, for the students* of the following faculties: Agriculture, Animal Husbandry, Veterinary Medicine, 3 months length (12 weeks) working with farms, the 1st series – 8 students in April – May – June 2001 and the 2nd series – 15 students in the same period , 2002. This practical training was offered by UASVM in collaboration with the Foundation for Rural Development and Ribe Amts Familielandbrug – RIFA Denmark, within a Leonardo da Vinci Programme.
- *Short-term practical training in Danish farms (3 weeks)* in order to help the Romanian farmers to learn more about the associative forms of organisation in the Danish agriculture assuring environment protection. In 2002, UASVM Bucharest, in collaboration with the Foundation for Rural Development and Agro-Forum National Centre for Agricultural Extension, Aarhus, Denmark offered 25 grants, financially supported by the Danish Ministry of External Affaires.
B. *Trainers' Training*
- *Agricultural University and high school managers (vice-rectors, directors)* attended a two weeks training course at Grinsted Agricultural College, Denmark, in 2002, within a Leonardo da Vinci Programme.
- *Intensive Training Courses for Consultants (ANCA and DADR staff) destined to improve their knowledge and skills* on: Farm Management, Agro-food products Marketing, The impact of C.A.P. In the year 2002, these 3 months length intensive courses have been attended by a number of 770 OJCA and DADR specialists from various counties
C. *Agricultural Extension Services, free of charge, for the farmers* operating in the area of DES and UASVM Bucharest, mainly from Ilfov and Călăraşi counties.
D. *Agricultural Extension services of high scientific level (paid services)* for Private Commercial Societies of high economic power (high turnover) from various counties. The main problems solved by consultants in the field are: Animal Nutrition, Animal Reproduction, Product Quality Control, Environment Protection.
E. *Short-term training courses for farmers*, organised within various programme financed by E.U. as follows:
- *For dairy farmers,* within a PHARE Project concluded between UASVM Bucharest and Reaseheath College, Great Britain, in collaboration with Cattle Breeders Association of Romania. In the period February – September 1998 and February 1999 a number of 80 farmers were trained. The topic of the courses was:
  o Dairy Cows Nutrition,
  o Milking Techniques and Hygiene,
  o Milk Quality,
  o Cattle Reproduction and Breeding,
  o Dairy Cow Health: mastitis and lameness,

- o Farm financial and business Management.
- *For crop producers and farmers dealing with animal production:*
  - o Within the W.B. Project Nr. 2240/12.XI 2001 (PMU, MAFWE) entitled "Extension of knowledge and skills in the field of agro-zootechnical farm management and business, agro-food processing and commercialisation" there were organised Short-term training courses. Target group: 400 farmers from IF, CL, GR counties. Training period: November 1st, 2001 – October 31, 2004. Topics: Farm Management, Marketing, Business Planning, Experimental Plots, Modern Technologies in Crop and Animal Production, Creation of Associative Forms of organisation for agricultural producers.
  - o Within the W.B. Project Nr.1801/2292/November 23, 2001 (PMU MAFWE) entitled "Integrated Informatics System for Business Planning in agricultural farms". Target group: 75 farmers from the following counties: IF, GR, CL, CT, BR. Training period: November 2001 – December 2004. Short-term training courses on the following topics: software improvement, courses and practical demonstrations about how to use the following soft wares: AGR-1 – for crop production; AGR-2 for animal production; AGR-4 for business planning.
- F. *Exchange of ideas and experience on agricultural extension* on the occasion of symposia annually organized by the faculties of UASVM Bucharest, where are invited to take part specialists from DADR, ANCA-CJCA, local councils, directors of agricultural commercial companies mainly from the counties situated in the area of influence.
- G. *Participation of UASVM academic staff* to various meetings with agricultural topics, organized by DADR in collaboration with the Producers Associations, OJCA: fairs, exhibitions of products and agricultural machinery, INDAGRA; on this occasion there are organized actions of consultancy, demonstrations, round tables etc;
- H. *Dissemination of academic staff research results* among the specialists, students, doctoral candidates, researchers, farmers by editing and publishing university textbooks, brochures, practical guide lines, papers in the university Bulletins and in various agricultural reviews such as: Romania's Agriculture, Agricultural Profit, the Farmer, Cattle Breeder's Review, Romanian Agriculturalist etc.
- I. *Agricultural Extension services for Milk Processors* within the PHARE – UASVM – Reaseheath College – Marea Britanie in the period February – September 1998 and February 1999. A number of 62 employees from Prodlacta Braşov Joint Venture Commercial Company attended courses in Milk Processing, Milk Microbiology, Analysis of Milk Quality, HACPP in milk processing.
- J. *Distance Learning* in Management, Economic Engineering in Agriculture and Rural Development, with two specializations: Management in Agriculture and Public Food and Agro-tourism, 6 years length, for professional and vocational training in the field.

An important role in the extension services delivered by agricultural universities play OZUCA - The Zone Offices for Agricultural Extension, founded within TEMPUS S - JEP 14051 - 1999 Project, entitled **"Development of Training Centres for Agricultural Extension at Agricultural Universities in Romania",** developed and implemented in collaboration with the following partners: University of Hohenheim, Germany, as contractor, UASVM Timisoara - co-ordinator, University of Wageningen, The Netherlands, UASVM Bucharest, UASVM Iasi, UASVM Cluj-Napoca, Faculty of Agriculture of University of Craiova, A.N.C.A.

OZUCA 's tasks and responsibilities:
- setting up modules for agricultural consultants training and specialisation (ANCA network Staff, private consultants or other specialised firms);
- assuring the infrastructure necessary for agricultural consultants specialisation, planning the length of courses and their timetable;
- assuring lecturers (academic staff), experts and specialists with responsibilities within OZUCA in collaboration with ANCA;
- editing and spreading informative materials needed for extension activity;
- identifying the priorities concerning farmers and consultants problems and creating a data bank;
- identification of partner institutions dealing with the same activities as O.Z.U.C.A. and creating opportunities to develop together common programmes and projects at national and international level;
- promoting OZUCA's image at national and international level;
- assuring optimum conditions as the proposed programmes of actions to be fulfilled, by a using in a corresponding manner OZUCA's funds;
- assuring special services for communication and information in agriculture.

Between the agricultural universities, ANCA and research institutes exists a close collaboration in running extension activities as follows:
- Co-ordination and harmonisation of the common programmes for agricultural extension;
- Joining the technical endowment used for extension activities;
- Assuring young graduates trained in Agricultural Extension to be employed within ANCA's network;
- Improving ANCA and DADRs specialists' knowledge in agricultural extension by organising postgraduate and master courses, short term training visits abroad, common training actions;
- A continuous bilateral informational flow between agricultural universities and the other institutions involved in agricultural extension.

Research institutes and stations involvement in extension services

Extension services are also delivered to dairy farmers by Research Institute for Cattle Growing and its 7 local research stations and Institute for Biology and Animal Nutrition. They have specific programmes for farmers training, demonstrations in the pilot farms, experimental plots, conferences, scientific sessions, participation to agricultural fairs, delivery of combined food, high value animals etc.

The main topics for farmers training are the following ones: how to produce high quality forages (silage, hay), ration balancing, cow feeding during pregnancy, dry period and at parturition, calf feeding, reproduction (insemination techniques, semen quality control, the best period for inseminating cows, how to assist cows at parturition, the best moment for re-inseminating a cow, treatments for uterus diseases), breeding (milk recording, selection criteria for bull mother, bull testing, select the best bull from a catalogue); cow and calf maintenance; milking (milking techniques; milking machines; milk hygiene); manure collecting and storage.

Research institutes put at the farmers disposal projects of model farms of various size: 10 cows, 20 cows, 30 cows, including all the required calculations concerning the investment. They also offer technical assistance for building a new farm or modernizing the old ones.

Researchers with long experience in dairy farming give concrete solutions to the problems coming from farmers any time when they are asked.

The research institutes deliver up dated information on scientific results to extension network, inviting consultants to take part to scientific sessions, symposia or to participate to research programmes.

They have also exchange of information with the agricultural universities, MAFRD, ADRD at local level and other organisms involved in agricultural extension for joining the efforts destined to help dairy farmers.

A. Role in explaining Government rules to farmers

Extension Services play an important role in explaining the Romanian Government and the EU Rules to farmers as follows:

- agricultural policies in the field of dairy farming and milk processing;
- government orders for supporting dairy farming with subsidies for milk production, obtained calves from AI, for buying high breeding value animals, per kilogram live weight in case of steer fattening, for assuring high quality certificated seeds etc;
- legislation as Land Law, Lease Law, Association Law, Sanitary and Veterinary Legislation for assuring animal health and milk and meat quality;
- the EU Rules have not been so much explained to farmers, but within the training sessions the farmers were warned that they have to be prepared for the moment when Romania will be accepted as an EU member, they have to pay attention to the following aspects: farm size, high performance technologies, milk hygiene, milk quality, training;
- consultancy and concrete help offered to farmers in order to learn how to apply for SAPARD, World Bank funds, according to the applicant guide lines;
- presentation of Associative Forms and co-operatives in Japan within JICA Programme of collaboration between A.N.C.A. and Japanese Extension Agency;
- Teaching farmers how to complete documents required for getting compensations for damages in case of animal illness;
- Information on farm inputs prices, but also for milk price on various markets.

B. Extension role in explaining farm management practices to farmers

Extension network plays an important role in explaining farm management practices to farmers, using various methods as follows;

- *individual methods*, according to farmers' concrete needs. For instance, how to set up optimized rations, how to treat mastitis, how to chose the highest breeding value bulls from a catalogue, how to produce a high quality hay or silage, where to buy and how to use the best way a milking machine etc;
- *group methods*, as follows:
- *training courses* for dairy farmers on the following topics: feeding, milking, buildings, breeding, reproduction; milk marketing; but also for the farmers with larger commercial farms on business planning; financial analysis; book keeping; costs, incomes and gross margin calculation; at the end of the courses dairy farmers receive a certificate, attesting qualification;
- *visits at model farms*, for example at Pantelimon Agroindustriala Commercial Company – the best private milk farm in the country;

- *meetings and lectures* delivered by invited speakers (researchers, university professors) in order to update knowledge in a special area or to solve a concrete problem of high difficulty; for example, how to use the AGR 1 software for optimizing forage production;
- *exchange of experience between farmers*: round tables, where the best farmers with high performance in milk production and turnover are invited to inform and teach the other farmers from their experience;
- *organization of demonstrative plots* in collaboration with research institutes such as: Institute for Research and Development for Cattle Growing and Institute for Biology and Animal Nutrition within the project entitled "Management of Dairy Farms situated in the proximity of the big cities – Bucharest".
- *visits at research institutes and stations* to see live how farm management is applied to get high performance;
- *demonstrations in the field* concerning forage producing (silage, using forage additives etc) in collaboration with nutritionists from Institute of Biology and Animal Nutrition and Pantelimon Agroindustriala Commercial Company;
- *demonstrations* in applied informatics, about how to use special software packages for optimizing crop structure, crop production, livestock structure, rations, costs etc (AGR 1 software); demonstrations are organized in collaboration with University of Agricultural Sciences Bucharest, Faculty of Management.

Training Courses for farmers are free of charge, are 3 - 6 length courses, being organized twice a week, late in the afternoon or every Saturday in collaboration with Research Institute Balotesti for Cattle rearing and fattening. The farmers receive a certificate attesting their knowledge and skills.

A.N.C.A and its network assures elaboration and editing of various publications of high interest to farmers: brochures, books, textbooks, magazines, newspapers, which are delivered for free to farmers, the related costs being covered by ANCA and MAFRD.

C. Urgent extension needs

In my opinion, the main extension needs at present are the following ones:
- reorganization of A.N.C.A. according to the 8 Euro-regions in the country in order to meet SAPARD Programme requirements in helping farmers;
- more high competence consultants so that each commune to have at least a consultant in dairy farming; at present there is only 1 consultant at every 5 - 6 communes;
- a better distribution of consultants specialized in dairy farming (zootechnicians) in the territory;
- all the consultants to be trained and get updated knowledge and skills in farm management within an accredited centre at national level; at present, it has been already signed an World Bank Project for 2 billion USD for creating the National Centre for Extension Officers Training within University of Agricultural sciences and Veterinary Medicine, Bucharest; the training will be delivered in collaboration with the other 4 agricultural universities in the country;
- new skills in Farm Management, Information Technologies, Communication;
- a better technical endowment within ANCA network: computers, cars for travelling in the field etc;
- substantial financial support from the EU and local Government funds for assuring a modern infrastructure, a higher salary, materials, fuel, experiences organized within pilot farms.

D. Extension needs when approaching the accession to EU

- trainers' training working with the National Centre for Agricultural Extension Officers Training – short training visits in the EU countries with a long experience in Agricultural Extension Services (UK, DK, The Netherlands, Germany);
- ANCA consultants training on Chapter 7 – Agriculture, from the Communautaire Acquis and Common Agricultural Policies;
- A continuous consultants education in Farm Management, based on the use of integrated information systems, IT, Communication, Agro-food Marketing in the EU countries; Milk Quota and Farm Management in the EU countries; how to become a consultant in the EU vision;
- A larger and comprehensive variety of topics to meet the uncovered niches in farmers' training as follows:
  - *for low trained farmers (lyceum, 8 classes):*
    - modern scientific and effective technologies for crop and animal production;
    - how to use modern infrastructure for agriculture (tractors, agricultural machines etc) for small and medium sized farms;
    - farm evidence, bookkeeping and management;
    - environment protection;
    - agro-food product marketing;
    - agricultural products processing;
    - integration in agriculture;
    - co-operation and association in agriculture;
    - legislation concerning subsidies for agriculture, loaning and taxation;
    - rural development.
  - *for highly trained agricultural specialists:*
    - modern farm/firm management in the market economy;
    - project management for developing agricultural production and services in the rural areas;
    - the use of IT in order to improve communication skills;
    - Romanian and E.U. agrarian legislation;
    - Common Agricultural Policies (CAP);
    - Sanitary and Veterinary Regulations provided by the Acquis Communautaire for Food Quality Control;
    - Agro-food products marketing;
    - Measures for environment protection and biodiversity preservation;
    - Modern biotechnologies;
    - Accounting and financial analysis for private farms and commercial companies;
    - Quality management according to the international standards;
    - Plant protection.

E. The agricultural extension system is facing the following problems:
- The lack of a corresponding legislation concerning agricultural extension: territorial organisation, financing, advisors formation and training, infrastructure, programmes management
- The lack of a permanent co-operation, co-ordination and communication between the bodies with responsibilities and tasks in the area such as: government, MAFRD and its ADRD network, ANCA and its CJCA network, agricultural universities, research institutes

and stations, SAPARD Agency, WB – MAFRD, Local Councils, other organisations (foundations, associations, etc);
- Ineffective extension services compared to similar advisory services provided in the countries with high developed agriculture;
- Low budgetary financial support to ANCA and its territorial network;
- The weak financial support offered by government to other organisms involved in agricultural extension (universities, research institutes, ADRDs);
- The agricultural producers are not able to support a private extension system from a financial point of view;
- Not sufficient (modest) infrastructure (lecture rooms, equipments for training and demonstrations, experimental plots, IT endowment, means of transportation in the field);
- Consultants' low salary, according to the budget system;
- The lack of an organised, modern and permanent framework - a special training centre for agricultural extension - where all the agricultural advisors to be trained;
- Farmers mentality, mainly of the poorest ones or having peasant households, concerning the importance of agricultural extension.

F. Extension services and other institutions to tackle the accession to EU

Romania's accession to the EU creates additional responsibilities to all the bodies involved in extension activities. Extension must be an effective tool to strengthen farmers 'knowledge and skills to produce more a high quality milk and to assure a satisfactory profit under milk quota.

The following problems which must be solved will be:
- In the field of legislation:
  o To change and complete the actual legislation concerning ANCA foundation and organisation in the sense to reorganise it by 8 Euro regions;
  o To include on the list of agricultural professions the one of „agricultural advisor / consultant" so that this new profession to be officially recognized;
  o Legal regulations providing that the consultant to give his/her approval for the agricultural works running in his area of activity;
  o Harmonisation of the Romanian legislation with the one existing in the E.U. countries concerning agricultural extension services;
  o Issuing an official document (law, Emergency Order, etc) concerning the foundation of a „Special Centre of Excellence for Agricultural Extension in Romania" with the seat within the UASVM Bucharest.
- In the field of organisation:
  o Setting up a National Centre of Excellence for training the specialists and farmers with the seat within the largest and the oldest agricultural university, UASVM Bucharest, which has the most numerous high competence staff and a modern infrastructure, the best in the country;
  o ANCA's reorganisation at territorial level in 8 regional centres corresponding to the 8 SAPARD Euroregions.
- In the field of consultants formation and training:
  o Assuring the trainers' training abroad in well known agricultural extension Centres from the E.U. countries;
  o Specialists' training in agricultural extension, according to the system existing in the E.U. countries, that is within a National Centre specialized, authorised and accredited in agricultural extension;

- o Setting up special programmes for trainers' training, with an aligned content to the E.U. requirements;
- In the field of extension programmes:
- o Common agricultural extension programmes developed in collaboration between universities, ANCAs network, DADRs, research network and other organisations;
- In the field of financing:
- o Setting up a strong financial resource system for agricultural extension, supported by the Romanian Government operating at the same time with the external E.U. and W.B. funds;
- o Assuring financing by destination (cost item):
  - Modern infrastructure in the territory by implementing the integrated information systems which will be helpful and useful for a faster and more effective communication between the organisms involved in agricultural extension and their beneficiaries;
  - Assuring a corresponding salary system for employees working within the extension service network;
- In the field of agricultural extension management:
- o Setting up the objectives and establishing the directions that extension activity has to reach;
- o Setting up plans of concrete measures destined to reach the proposed objectives;
- o Establishing the tasks and responsibilities;
- o Identifying the financial resources;
- o Assuring infrastructure at the level of the present requirements;
- o Consultants recruitment, training and promotion;
- o Establishing the modern extension techniques and methods that have to be applied in the territory;
- o Establishing formal and non-formal communication channels between extension advisers and their clients;
- o Establishing the corresponding criteria for extension activity assessment;
- o Setting up a data bank about agricultural extension in Romania.
- training programmes for dairy farmers on the topics of high interest (see point E)

The biggest challenge for the dairy industry in the coming years

The biggest challenge for the dairy industry in the coming years is MILK QUALITY which must be assured along the whole milk chain according to the EU standards.

What the dairy industry has to do to cover the requirements? It surely has to pay a special attention to the following aspects:

- a severe sanitary veterinary control of all the dairy farms, selecting the ones with a chance to correspond the requirements imposed by milk quality;
- milk quality control within the specialized laboratories concerning number of pathogenic germs and somatic cells, besides fat and protein percentage, acidity, etc.; concerning this aspects, the laboratories must be modernized and well endowed at EU standards; they have to issue milk quality certificates;
- milk quality standardization according to the EU criteria till the end of December 2009, accepted by the EU Commission at the last negotiations in June 2004;
- a new legislation concerning milk producing, collecting, storage, transportation, quality control, aligned to the EU legislation;
- a severe milk reception at the level of milk processors and in the free markets;

- implementation of quality management system (HACCP, etc) within milk processing factories;
- establishment of milk collection points endowed with milk tanks and special equipments for assuring a corresponding milk storage;
- milk transportation means under corresponding hygiene and microclimate conditions;
- special storage conditions for dairy products;
- collecting and processing biological milk in the farms situated in the hilly and mountain areas; biological dairy products have a chance to be sold on the external markets at a higher price;
- modernization of infrastructure within milk processing factories; it was accepted by the EU Commission for a 3 years period till December 2009.

References

Gavrilescu, D., 1994. Durability of Agricultural Family Farm – Anticipated Trends, Research Institute for Agricultural Economics, Bucharest, Romania.

Gavrilescu, D., 1995. A chance of Family Farm in Romania. Quality of Life, No 3-4, Romanian Academy Press House, Bucharest, Romania.

Popescu, A., S. Baiko & E. Beck, 1999. Financial Analysis in Brasov Dairy Farms. Cattle Breeders Magazine, Bucharest, Romania.

Popescu, A., E. Beck & S. Gyeresi, 2000. A comparative study on Production, Vosts, Incomes and Gross Margin in some private dairy farms. Achievements and Foresights in Animal Husbandry and Biotechnologies, National Symposium, Cluj-Napoca.

Popescu, A., 2001a. Research concerning the Financial Results and Risk factors in Dairy Farming. Foresights on Restructuration and Recovery of the Romanian Agriculture. National Symposium, Bucharest, Romania.

Popescu, A., 2001b. Research on Economic Results of Private Dairy Farms. International Symposium "Prospects for the 3rd Millennium Agriculture", Cluj-Napoca.

Popescu, A., 2003. Financial Management and Business Management in Dairy Farms, Agris Publishing House, Bucharest, Romania

Romanian Yearbook 2003, Romania.

National Strategy for Durable Development of Agriculture and Food "Horizon 2025", Ministry of Agriculture , Forestry and Rural Development, Romania.

Farm management and extension needs in Bulgaria

Elitsa Zdravkova

Ministry of Agriculture and Forestry, 55 Christo Butev Bulv, 1040 Sofia, Bulgaria

Introduction

Bulgaria is a country with suitable climatic and natural conditions for dairy farming, which has always been a traditional agricultural sector. The dairy sector is also very important to the economy of the country, since the major number of bovine animals are dairy cows. The average share of the gross production of the dairy sector in the gross agricultural production is about 12%, and the average share of the sector in the gross value added is 27%. Bulgaria is also a traditional exporter of Bulgarian white brine cheese and curd, which amount on average to about 4% of the total agricultural exports in the last few years.

The consumption of milk and dairy products in Bulgaria is traditionally high, even though in the beginning of the last decade it experienced a severe drop of about 40%, due to the economic crisis and the decrease in income that went together with it. Over the last 3 years the tendency has been for a gradual recovery. One of the key products of the Bulgarian dairy sector is Bulgarian yogurt (sour milk), which has no equivalent in the world and with its superior taste and health benefits, it could fill in a great market outlet.

Figure 1. Map of Bulgaria.

Current situation and structural changes in the near future

The dairy sector in Bulgaria is currently very dynamic and is changing literally as we speak of it. Therefore, and due to the fact that no thorough survey has been made of the sector, it is very hard to make accurate predictions for the next decade. The structure of the dairy sector at the moment is exemplified by the following Table 1.

Table 1. Number of dairy cows and distribution of cows in the dairy farms.

Number of dairy cows/farm	Number of farms	% of all farms	Number of dairy cows	% of all dairy cows
1 cow	130 907	68,6	130 907	36,2
2 cows	39 320	20,6	78 641	21,7
3 – 9 cows	17 082	8,9	72 387	20,0
10 – 19 cows	2 385	1,3	29 388	8,1
20 – 49 cows	973	0,5	26 131	7,3
50 – 99 cows	185	0,1	12 084	3,3
Over 100 cows	78	0,1	12 309	3,4
Total number	190 930	100,0	361 847	100,0

As is obvious from the table, the predominant number of farms at the moment own 1 or 2 cows, and there is a significant number of farms with 3 to 9 cows. There is a slight tendency for smaller farms to unite or fall out of business (statistics show that there are about 6 – 8% less small farms every year), and for bigger farms to grow on the one hand and increase in number on the other. We expect this tendency to become stronger, so that the already bigger farms can better face the challenges of the market and EU accession.

The average production per recorded cow is shown in the Table below, and the tendency is towards higher figures in the next 5 to 10 years, due to the selection programs, carried out on the bigger farms.

Table 2. Average milk yield per recorded cow, by breed (2003).

Breed	Milk production (kg)	Fat content (%)	% of all controlled dairy cows
Holstein Friesian	4,853	3.75	74.0
Brown Swiss	4,253	3.91	17.9
Red Holstein	5,296	4.11	1.0
Simmental and cross-breeds	3,924	3.95	2.5
Rhodope short-horn cattle	3,795	4.71	3.6

In 2003 the number of dairy cows on individual selection controls has increased with 5.2% (1220 cows) and has reached 24 456, which is 6.8% of the total number of cows in the country.

The number of dairy factories is 295, of which 23 are licensed to export to the EU (data as of 2 July 2004). There are 3 627 milk collection centres.

However, those average figures are not very telling about the situation in the whole country. There are substantial regional differences in the distribution of big and small dairy farms, milk collection centres and dairies (see also annexes).

Table 3. Regional distribution of dairy cows, small dairy farms, dairies and milk collection centres (2003).

Region ([12])	NW	NC	NE	SE	SC	SW	Total
Number of cows	25,729	55,502	76,936	36,733	127,266	61,089	361,847
Number of cows in farms with 1-2 cows	13,384	24,968	39,246	17,672	80,522	33,017	109,548
Milk collection centres	428	807	1,070	217	910	95	3,627
Dairies	13	58	60	26	86	52	295
Dairies licensed to export to the EU	1	1	6	5	8	2	23

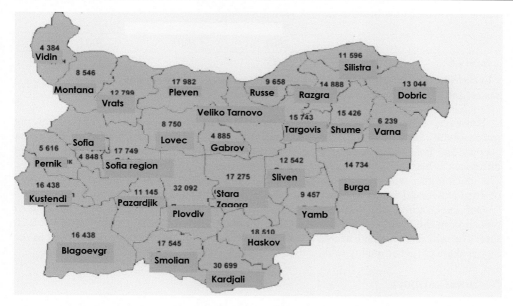

Figure 1. Number of dairy cows (by region).

[12] NW (north west) includes the administrative regions of Vidin, Montana and Vratsa.
NC (north central) includes the administrative regions of Pleven, Lovech, Pleven, Gabrovo, Veliko Tarnovo and Russe.
NE (north east) includes the administrative regions of Dobrich, Varna, Shumen, Silistra, Targovishte and Razgrad.
SE (south east) includes the administrative regions of Burgas, Yambol and Sliven.
SC (south central) includes the administrative regions of Plovdiv, Pazardjik, Stara Zagora, Haskovo, Smolian and Kardjali.
SW (south west) includes the administrative regions of Sofia city, Sofia region, Pernik, Kustendil and Blagoevgrad.

We expect more specialized dairy farms in the future as opposed to mixed farms. It is the practice that most dairy farms produce their own forage, but only for their own needs, and not for sale – we expect this to continue, since this is the cheapest way to get forage. Part-time farming shows a downward tendency of 6-8% less such farmers every year, and those are mainly the small farms. We expect this trend to continue and to have less and less part-time farmers in 5 years.

We also expect extensification in dairy farming, for a number of reasons:
- the yield per cow shows an upward trend, due to selection,
- the quota, which will be a limiting factor for the production of milk,
- the transfer of quantities from on-farm consumption to deliveries and/or direct sales.

Therefore, since we can only produce and market a specific amount (979 000 tons) and the on-farm consumption is expected to go down together with the declining number of part-time farmers, and the yield per cow is expected to keep going up, we will not need as many animals to produce the same amount of milk as we do now. Therefore, extensification seems like the only reasonable option.

Meanwhile, the average purchasing milk price has been showing a slight downward trend over the last 3 years, mainly due to the increase in production.

Table 4. Average milk price in various years (per litre).

	2000	2001	2002	2003
In BGN	0.34	0.35	0.33	0.31
In Euro *	0.1738	0.1789	0.1687	0.1585

Source: AMIS (Agro market information system);   * 1 Euro = 1.95583 BGN

However, serious differences can be noticed between the prices in milk from big (professional) farms, with over 10 - 20 cows (which represent about 2% of all farms in the country and hold about 22% of all dairy cows) and small dairy farms. The other very big difference is between summer prices and winter prices, the difference being up to two times. The much higher winter prices can be explained with the bigger costs incurred (greater share of dried and mixed fodder as opposed to the grazing period) and the smaller production.

Our expectations are that with the introduction of the quota system and the gradual reduction in the quantities of milk marketed, the milk price will reach a level of about 0.20 Euro and will remain stable at that level. This increase in the price will be mainly due to the bigger share of higher quality milk.

National support

We currently have subsidies for the milk sector (including sheep, goat and buffalo milk) and suckler cows support. No beef premiums are available to the farmers as yet.

The milk subsidies are given for high quality milk only, and the total sum of support for the cow's milk sector in 2003 was 8 million BGN (4.090335 Euro).

The suckler cow premium covered 36,000 suckler cows in 2003, which were included in the national fund. Aid was also given for the import of elite breeding animals.

Subsidies are also given to the National Association of milk producers for carrying out of individual controls (which was done for the first time in 2004). This shows that the government is cautious in performing tasks that are private in their nature, while at the same time supporting the associations in their first steps to implement these tasks.

There are also export refunds given for the export of dairy products to third countries (1.2 million BGN).

Ad hoc aids are also given for supporting the farmers with the price of the forage – this was the case in the winter of 2003-2004 (December 2003 – February 2004) and the aim was to compensate the farmers for the increased price of forage. This aid was given to registered agricultural producers and was on the basis of number of animals owned. The program covered 90 000 dairy cows and 138 000 sheep.

A budget of all the financial support is given in the Table 5, and it is obvious from it that the levels of support are constantly going up.

Table 5. National support for the dairy sector (BGN).

	2002	2003	2004
Subsidy for the national genofund			
1. Bovine animals	421 800	1 502 400	1 872 200
2. Sheep and goats	828 760	662 700	681 200
Subsidy for import of elite animals			
1. Bovine animals	45 000	--	331 000
2. Sheep and goats	--	235 000	125 000
Subsidy for high quality milk	2 300 000	6 000 000	8 000 000
Subsidy for forage	--	1 356 300	4 386 700
Subsidies for breeding associations in the milk sector	--	--	103 300
Export refunds	--	--	1 200 000
Total	3 595 560	9 756 700	16 699 400

Farm management

The largest cost factor is definitely forage, which makes up about 70-80% of the total expenditure in a typical dairy farm. Hence the big share of farms that typically grow their own forage (the price can then go down to about 65%).

Organization of extension and extension needs

In the fall of 2003 and the spring of 2004 the MAF started an information campaign, through which to disseminate information on the market mechanisms under the CAP that the farmers would face upon accession.

The main extension service provider is the National service for agricultural advice (NSAA), which is a part of the MAF. It has 28 regional offices, and provides all kinds of advice to farmers.

The MAF also has 28 regional directorates (RDAF) and 246 municipal services, which provide advice and support to farmers at the regional and at the municipal (the lowest) levels.

This year the MAF has also started a vocational training of farmers, under a SAPARD project.

All the aforementioned campaigns, services and advice are absolutely free of charge.

The last institution that provides extension services will be the National Milk Board, whose preparation is under way – the legal basis is already in place, and the preparation of its rules of procedure is under way. It is envisaged that this will be the body competent for the administration of the individual quotas, while the national quota and the reserve will be managed by the MAF.

As mentioned above, the main task of the information campaign that started in the fall of 2003 was to disseminate information on the market mechanisms under the CAP that the farmers would face upon accession. This included information on what common market organizations exist, what products they cover and what mechanisms are included. The farmers were also acquainted with the measures that the MAF has envisaged to implement with respect to accession, and their own responsibilities and respective rights to premia, payments, entitlements, etc.

The same set of information was also transmitted to employees from the RDAFs and the NSAA, so that they are better able to provide consultations to the farmers with regard to EU matters.

As mentioned above, this year the MAF has started a vocational training of farmers, under a SAPARD project, which focuses on farm management practices. Overall advice concerning farm management is also provided by the NSAA.

Challenges for the dairy industry in the years ahead

There are many challenges for the dairy sector, both connected with the restructuring of the sector itself and the upcoming accession to the EU. The biggest challenge, however, remains the establishment of a quota system – although we have started the process already, we are still in the very initial phases and we still have a long way to go and many unanswered questions, such as:

- the establishment of objective criteria for the allocation of individual quota (so that the quota system is introduced in a fair and correct way);
- whether we should establish regional quotas or not;
- whether we should allow quota to become an asset (to be bought, sold, leased) ([13]), and if yes, how will quota be marketed (through a special quota market, through the reserve, directly);
- whether buying and selling of milk should be geographically confined within certain regions, so as to avoid the dislocation of quota from one region to another ([14]).

These and other difficult questions (some purely technical, other more political) still lay ahead of us. Therefore, we believe that we have a lot to learn from what other countries have experienced and achieved so far.

---

[13] We are receiving advice in different directions and we fully realize that although this might hasten the restructuring process, it might also lead to misuse of the quota system.

[14] This is a very sensitive issue, since we have two large dairies (the French Danone and the German Meggle), which buy milk from all over the country, while the other smaller dairies buy the milk on a more or less regional principle.

Prospects of quota and farm management in Croatia

Kresimir Kuterovac

Croatian Livestock Center, Ilica 101, 10000 Zagreb, Croatia

Introduction

Cattle production in the Republic of Croatia has been in stagnation and decline for a longer period of time. The main reasons for the decline in the number of cattle and milk production has been the dismantling of large agricultural enterprises in the early nineties and the war on the territory of the Republic of Croatia. At that time heavy disorders inflicted the structure of cattle breeding, so cattle production has barely started to recover in the last few years. The structure of production is extremely unsatisfactory. Old, small and mixed-type family farms make the main part of production; low number of animals (3 cows) mostly lower levels of technology affect the level of production. There are some larger mixed-type family farms that use improved technology, that allow rather high levels of production. We also have specialized dairy farms that are on high technology levels and they also achieve high levels of production. In the future, it is necessary to restructure the present cattle production and use production systems which will enable the implementation of a high technology and productivity level and provide self-sufficiency in milk production.

General information

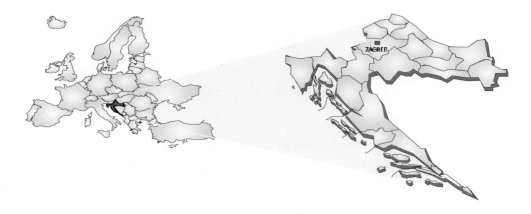

Figure 1. Location of Republic of Croatia in Europe.

Table 1. General data.

Total surface area of the Republic of Croatia, km^2	58,542
Agricultural surface area, ha	3,156,000
Population	4,282,000
Agricultural surface area per capita, ha	0.34
Brutto national income, $	5,056
Total unemployment rate, %	22
Foreign trade balance, mio. $	5,815
Spreading rate of import and export, %	45.72
Value of agricultural products in brutto national income, %	9
Employment rate in agriculture and forestry, %	16.5

Table 2. The structure of agricultural land.

Category	Family farms (km^2)	Enterprises (km^2)	Total (km^2)
Agricultural surface	2,089	1,067	3,156
Cultivated fields and gardens	1,154	303	1,457
Pastures	464	698	1,162
Meadows	350	58	408
Orchards	67	3	70
Vineyards	54	5	59

Table 3. Review of farms by size.

Size of a farm	Number of farms
Less than 1 ha	227,434
1 - 3 ha	112,062
3 - 5 ha	45,732
5 - 10 ha	42,426
>10 ha	20,878

The number of cattle per hectare is 0.34 and the number of cattle per capita is 0.10.

Milk production balance

Table 4. Cattle production in the Republic of Croatia in years 1990 to 2002.

Year	Number of cattle			Annual increase (tons)	Milk production (000 lit.)	Milk production per cow (lit.)	Redemption of milk		
	Total	Cows & pregnant heifers	Cows in lactation				Total (000 lit.)	Per cow (lit.)	Marketa-bility (%)
1990	830,000	492,000	460,000	117,000	889,000	1,932	342,273	743	38.49
2000	438,422	287,000	254,702	62,908	587,634	2,307	380,000	1,491	64.62
2001	417,113	277,668	254,293	56,387	634,519	2,495	409,329	1,609	64.51
2002	444,000	280,000		68,402	674,767		444,350		65.85

Source: State statistic institute

Total production in year 2002 was 674 mil. litres of milk. Since 1990 the number of cattle has been extreme decline, consequently the production and redemption of milk.

Production-consumption balance of milk in the Republic of Croatia

Table 5. Production-consumption balance of milk in the Republic of Croatia (2001 and 2002).

		2001	2002
	Cows	635,000	660,000*
Milk production in 000 lit	Sheep and goats	18,183	20,000
	Total	653,183	680,000
Feeding (3%)		19,596	20,400
Losses (4%)		26,127	27,200
Available milk in 000		607,460	632,400
Import - export	Import lit.	196,600	203,001
(milk equivalent)	Export lit.	75,100	63,486
	Balance lit.	121,500	139,515
Total available milk in 000 lit.		712,050	753,315
Population – Republic of Croatia		4,437,460	4,437,460
Provisory inhabitants (foreign tourists)		104,109	104,109
Industrial processing of milk, %		70.29	67.07
Marketability of milk, %		65.45	68,28
Self-sufficiency, %		83.33	81.92
Consumption	From domestic production	137.52	139.28
(litres/head)	From import	26.75	30.72
	Total	164.27	170.00

Source: State statistics institution;   * Evaluation of milk production for 2002.

The Republic of Croatia is net importer of milk and dairy products, and the percentage rate between import and export was 41.64% in 2002. At present the consumption is 170 litres per head, thus Croatia lacks about 150 mil. litres of milk, which is about 20% of total milk production in the Republic of Croatia.

The structure of cattle production

Table 6. Number of cows and breeders.

Year	*Total number of cows	Suckler cows*	Number of cows in milk recording			
			Family farms	Enterprises	Total	Total in %
1993	241,997	-	34,943	5,613	40,556	16.76
1994	226,231	-	39,051	5,178	44,229	19.55
1995	235,400	-	42,689	4,455	47,144	20.03
1996	233,477	-	48,247	4,132	52,379	22.43
1997	233,207	-	62,205	3,602	65,807	28.21
1998	230,650	-	75,921	6,224	82,145	35.61
1999	228,014	-	80,198	6,218	86,416	37.90
2000	214,666	-	85,459	7,439	92,898	43.27
2001	219,782	-	91,235	7,606	98,841	44.97
2002	224,078	-	101,157	7,367	108,524	48.43
2003	223,954	65,155	133,064	6,895	139,959	62.49

Source: Vet. stations;   *65,155 = Number of cows who are not under selection and government support.

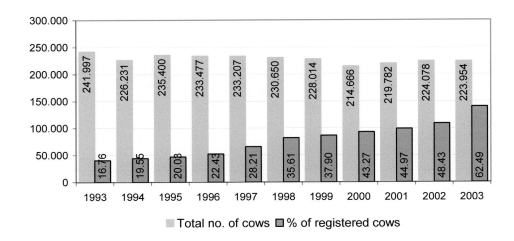

Figure 2. Total number and % of recorded cows.

The number of recorded cows in herd book is constantly increasing as well as the number of cows in milk recording. In 2003 there were 62.49% of cows in herd book register, and 20% in milk recording.

Table 7. The structure of milk suppliers by quantity classes.

Litres	Number or producers	% of total number of producers	Delivered quantity of milk (litres)	% of total delivered quantity of milk
6,000	34,110	58.00	89,697,226	17.01
6,000 - 10,000	10,452	17.77	80,893,593	15.34
10,000 - 20,000	9,262	15.75	128,039,150	24.28
20,000 - 30,000	2,613	4.44	63,187,158	11.98
30,000 - 40,000	1,043	1.77	35,804,354	6.79
40,000 - 50,000	478	0.81	21,317,382	4.04
50,000 - 60,000	270	0.46	14,722,069	2.79
60,000 - 70,000	176	0.30	11,374,516	2.16
70,000 - 80,000	114	0.19	8,473,267	1.61
80,000 - 90,000	65	0.11	5,499,956	1.04
90,000 - 100,000	41	0.07	3,888,231	0.74
>=100,000	191	0.32	64,446,286	12.22
Total	58,815	100.00	527,343,188	100.00

At present there are about 100,000 family farms in milk production business, and 58,815 of them deliver milk on the market. Most of them are small producers, so 58% of all dairy farmers produce less than 6,000 litres, and they deliver about 17% of total milk volume in 2003.

Table 8. Farms by size and the number of cows.

Size of a farm (ha)	Number of farms according to the number of cows				Total number of farms
	1	2 - 5	6 - 10	>10	
do 3.0	865	25,310	668	177	27,020
3.1-5.0	289	25,996	1,776	146	28,207
5.1-8.0	140	23,710	6,171	269	30,290
>8.0	64	11,908	9,138	2,014	23,124
Total	1.358	86,924	17,753	2,606	108,641

Source: Calculated from the register of statistic specification 1991, and the de facto situation 1998.

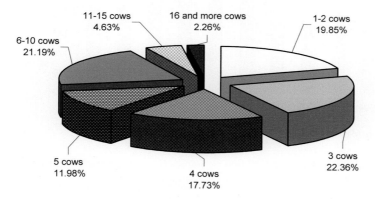

Figure 3. Breeders structure by herd size.

Small farms with small number of animals per farm and undersized agricultural surface area per farm predominate the structure of family farms.

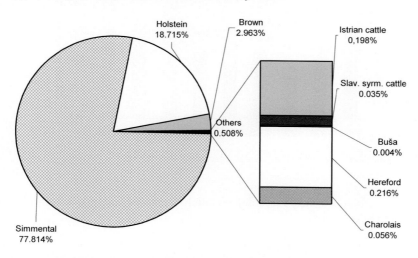

Figure 4. Breed structure of cows in milk recording.

The Simmental breed represents the largest number of the cows in milk recording (77.8%), while the Holstein-Friesian breed accounts for 18.7% of the cow population under control.

Milk production (standard lactation) of cows under recording

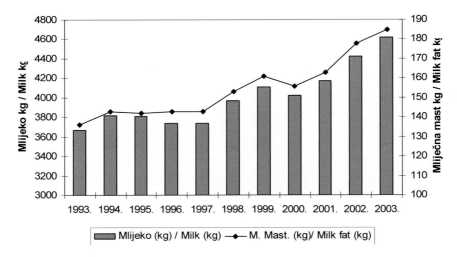

Figure 5. Milk production in standard lactation on family farms and enterprises from 1993 to 2003 for Simmental breed.

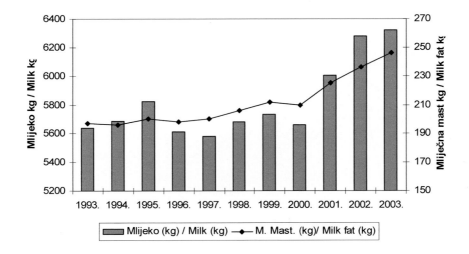

Figure 6. Milk production in standard lactation on family farms and enterprises from 1993 to 2003 for Holstein-Friesian breed.

The production of cows under milk recording is increasing, primarily because of better utilization of genetic capacity, and due to the results achieved by breeding and selection work.
Costs of milk recording procedure

In Croatia, milk control is carried out by AT method, and we also have a small part of the herds (20%) under B control. The cost of milk recording is 120 kn per cow per year (50 litres of milk). Breeders pay 50% of that cost, while the other 50% is paid by the government. Besides the analysis on milk fat, protein, lactose content and the number of micro-organisms and somatic cell count, we also control the urea content in milk. Milk recording is performed according to the ICAR recommendation. Croatian Livestock Center has got the right to use the ICAR special stamp.

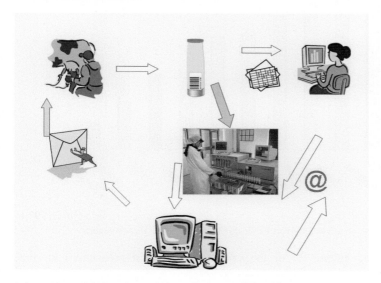

Figure 7. Milk control on the family farms in Croatia.

Government support in milk production

Milk producers receive government support for crop production, milk production per litre, and for keeping and breeding of cattle.

Table 9. Type of subsidy in Croatia.

Type of production	Type of subsidy	Amount (kn)
	1 ha arable land	1,650
Plant production	Forage crops	1,250
	Meadows and pastures	100
	Milking cows (annually per head)	800
Animal production	Suckling cows (annually per head)	1,500
	Calves (female)	1,500

Subsidy for cows milk is calculated depending on quality and it amounts to 0.40 - 0.75 kn/l. Subsidies are applied also in sheep and goat production and for sheep and goat milk, which is 1.0 kn/l. Total subsidy on milk production in the Republic of Croatia amounts to about 35% of the total production value (Figure 8).

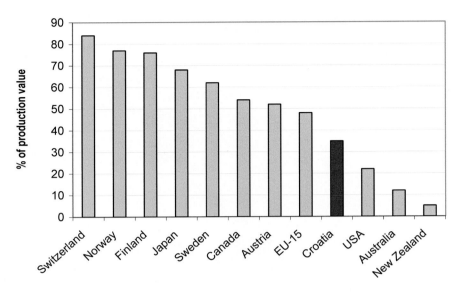

Source: Preposition of Development strategy of agriculture for the territories of special government custody

Figure 8. Subsidy for agriculture (% of the value of agricultural production) in some countries.

Farm management

The issues of farm management arise primarily from the family farms structures. Small family farms are not organized and they lack the knowledge and skills to increase the level of production per cow. Current production is 3,300 litres per cow. Farming economy restrictions are also poor condition of housing, care and feeding. The perspective of small family farms is poor also due to the age structure of the owners.

Table 10. Costs of milk production (per cow/year).

Costs	Kuna*	%
Feeding	8,609.78	79.26
Veterinarian costs	1,080.00	9.94
Breeding costs	73.20	0.67
Bedding	400.00	3.68
Other costs	700.00	6.45
Total	10,862.98	100.00

Calculation catalogue HZPSS 2004. (Croatian Institute of agricultural consulting service)
* 1 euro = 7.39 Kuna

Family farms owned by younger and more ambitious owners can accomplish development of cattle production on their farms. It is necessary to conduct systematic education of young future farmers and permanent education of all employees in the milk production segment, so that they are able accept new technologies and meet the requirements of new standards in milk production. In the future, the priority will surely be the improvement of production technology and application of new technologies, better feed utilization as well as improvement in economics and savings in milk production.

Table 11. Purchase price of milk in Croatia - 03/2003 (4.10% milk fat and 3.35% protein).

Elements – price		Class			
		E	I	II	III
Main price/kn		1.9370	1.937	1.9370	1.9370
Add in – subtraction for class/kn		+0.2697	0.000	-0.0899	-0.1798
Dairy subsidy/kn		0.1200	0.120	0.1200	0.1200
Purchase price of dairies	kuna	2.3254	2.056	1.9658	1.8759
	€	0.3100	0.280	0.2700	0.2500

Source: Professional services;   € = 7.39 kn

The price of milk in the Republic of Croatia is at the same level as in the countries of the European Union. It is related to the class of milk which is calculated according to the number of micro-organisms and somatic cell count. Regulations on fresh milk quality are in accordance with EU directive and have been implemented in the Republic of Croatia since January 2003.

Milk quality control in the Republic of Croatia is organized through a central laboratory for milk control. The results are used for payment purpose, and for breeding, selection and management activities.

The new system of milk quality control has been introduced in 2002. It includes the unique database of all milk producers, organization of sample collection, transportation of samples, sample analysis in the laboratory, feedback to milk producers, dairies and competent inspection.

Organization of advisory services

- There is a specialized governmental institution in the Republic of Croatia whose main activity is providing services and education for agricultural producers. Its service is organized through regional offices, available to all interested producers. Its employees are agricultural engineers, veterinarians and economists. The main restriction of institution's functioning is the limited number of employees and the modest budget available. The service has recently accepted new educational methods and approaches. The costs of activities of services are completely paid by the government. There is another institution that acts in the livestock breeding segment called the Croatian Livestock Center. It is predominantly a control institution which, because of the nature of its activity (control of production traits of domestic animals), carries out also a specific part of advisory services.
- Institutions have an important role in explaining the new rules to farmers, they also give farmers all the necessary information and organize education through which they help producers in making the necessary changes in their production. One of the recent examples is introduction of the new milk control system. The whole system was set up in a short period of one year. The main problem was to inform producers with the implementation of the new milk control system, and the need for introduction of new standards in milk production and improvement of milk quality.

The most important tasks of advisory service are to inform milk producers with the future conditions of milk production, new standards and new milk production technology.

It is necessary to provide systematic organization and education for farmers, and the local advisory offices have to help in the restructuring process of family farms and production specialization. It is also necessary to provide an additional quality services in the segment of farm management. The advisory services should focus their activities to farmer development and organization.

That is exactly what is expected in the process of joining the European Union – expert services should be well organized and equipped with new possibilities, to provide high quality services to milk producers and to be highly qualified for the quality education of farmers. All our institutions will have to adjust to the standards and demands of the European Union in carrying out of their activities. Adjustments are needed in the activities they conduct, and the main goal will be a creation of modern, flexible professional institutions which will conduct their activities according to the standards of the European Union.

In doing so, the efficiency of all the involved institutions and the general condition in agriculture must be taken into the account in the Republic of Croatia.

Which are the challenges in the future of milk production?

In the coming period intensive investments in cattle breeding and milk production are planned in the Republic of Croatia to arise the current production to the level which will provide self-sufficiency of cattle products and their competitiveness on the EU market.

It is necessary to restructure some of the current farms and build up some new production units. The rest of small production units, with no possibilities for adjustment, will stay as long as production is continued by the farmer who is often at older age.

The new programme of cattle breeding development has been accepted by the Government of the Republic of Croatia. It is based on minimal increase of the number of cows, changes in family farm structures, and maximal exploitation of the available agricultural land.

Table 12. Development of milk production in years 2001 – 2011.

Attribute	Years	
	2001	2011
Milk production (000 lit)	634,519	1,071,161
Milk purchase (000 lit)	409,329	915,485
Farm size (ha)	2.90	
Farm capacity (cow)	2.80	10.8
Milk production per farm per year (kg)	6,000	42,846
Milk production (per cow in litres)	2,495	4,217
Milk sale per farm (litre)	5,738	36,619
Milk sale (per cow in litres)	1,747	3,604
Milk marketability (%)	64.51	85
Self-sufficiency (%)	83.33	90
Consumption (lit/p.c.)	164	220

Source: State statistic institute, Ministry of agriculture, forestry and water management

Farm structure will change through the anticipated Programme: a decrease of small farms will occur and the number of cows will be reduced on such farms, while at the same time the number of specialized farms will increase, as will the number of medium size mixed-type farms.

In the future period, we expect intensified milk production, and an increase in cattle number per hectare.

Part time farmers will surely represent a significant part of the milk producers. Despite of all expected and planned changes, the structure will not be able to substantially move towards specialized dairy farms because of the number of limitations that exist, mostly regarding the use and availability of agricultural land.

Table 13. Evaluation of farm structure and number of cows during the application of cattle production program.

Farm group		Farms number during the application		Cows number during the application	
		Beginning	End	Beginning	End
Small	Existing	96,300	42,300	279,200	169,200
	New	-	-	-	-
	Total	96,300	42,300	279,200	169,200
Adapted	Existing	400	5,700	4,800	84,850
	New	500	300	6,000	6,000
	Total	900	6,000	10,800	90,850
Specialized – family farms	Existing	20	1,100	700	43,500
	New	80	100	2,800	4,500
	Total	100	1,200	3,500	48,000
Specialized – enterprises	Existing	15	19	6,750	11,400
	New	1	1	450	600
	Total	16	20	7,200	12,000
Total – dairy		97,316	49,520	300,700	320,050
Suckling herds	Existing	-	213	-	12,990
	New	10	15	500	1,050
	Total	10	228	500	14,040
Total farms		97,326	49,748	301,200	334,090

Cattle production development Program in the Republic of Croatia (2001-2010).

Milk price will not change significantly according to the estimation of dairy industries, so that domestic industry will not be in an unfavourable position compared to the EU countries.

Government support will change within the meaning of type of subsidy. The total amount of subsidy for milk production will not change significantly. It should not decrease, because the funds are necessary for successful accomplishment of the ambitiously prepared plan for the renewal plan of cattle breeding in the near future.

The impact of programme on milk production

Table 14. Estimation of milk production during the application of the programme.

Farms group	Application period			
	Beginning		End	
	Milk production		Milk production	
	Per cow	Total	Per cow	Total
Small	2,300	642,160,000	2,500	423,000,000
Adapted	3,500	37,800,000	4,500	405,000,000
Specialized – family farms	5,000	17,500,000	6,000	288,000,000
Specialized – enterprises	6,000	43,200,000	7,000	84,000,000
Total	2,466	740,660,000	3,604	1,200,000,000

Cattle production development program in the Republic of Croatia (2001-2010).

The main challenges in milk production

Application of cattle production development programme in the Republic of Croatia represents a main challenge in cattle production. The realization of this programme in a short period will have a long-term effect on the agriculture of the Republic of Croatia. Cattle production provides a significant rate of employment for a large number of family farm members. On the other hand, the Republic of Croatia will, by joining the EU, get quota for milk production which will affect our cattle breeding and milk production volume for quite a long time.

Figure 9. A modern dairy farm in Croatia.

Figure 10. Cattle shows are part of agricultural life in Croatia.

Farm management and extension needs in Albania

Niazi Tahiraj

Ministry of Agriculture and Food, Bulevard Deshmoret e Kombit, Tirana, Albania

Livestock production tendencies

Traditionally agricultural production, has provided the most important part of the Albanian economy. The contribution of agricultural production to **G**ross **D**omestic **P**roduction (GDP) was 50% in 2000-2001, 37% in 2002 and 34% in 2003. It was reduced because of the diversification and industrialization of the economy, structural transformation inside and among sectors of economy.

Livestock production growth rate is high compared to the other sectors of agriculture, so the development of livestock production will continue to be a very important component in Albanian agriculture. It represents 48% of agricultural production. Actually, the Albanian livestock products are sold predominantly on the domestic market and they cover about 99% of milk, eggs and honey needs, and approximately 60% of meat needs. What has significantly changed is the rate between domestic production and imports. The domestic production has increased and the importation of milk, eggs and honey has reduced to a minimum. Importation of meat has also decreased.

During 2002-2003 period, we completed the Report on "The Animal Origin Genetic Resources Situation in Albania", which was prepared within the frame of World Report - guided and directed by FAO.

The report served our Ministry of Agriculture and Food to conduct the right policy of good management in livestock production and to protect the local breeds being at risk of extinction. So, recently, buffalo local breed has been taken under the Albanian government special protection to escape the extinction.

The Government policy is to encourage the local business aimed at liberalising and reducing the tariffs of importation of breeding material as well as feeding raw material. The establishment of combined livestock and industrialized food production industry was stimulated through the policy of price reduction for raw material of protein nature, which is imported from other EU countries.

Increasing trend in livestock production was evident in the early nineties, but in 1997, the number of animals was reduced, while the number of sheep and poultry increased in 1998. Low increasing rhythm was characterized for pig production sector.

Livestock structure

Cattle production is the most important sector of livestock production which provide 85% of milk and 53% of meat production. Cattle are bred in about 320,000 private family farms.

The predominant cattle breeds are: Jersey cows and crosses account for 42% of cow population, Black and White breed and crosses reach 38%, *Tarantaise* breed and the crosses 2%, *Oberintal* and crosses 2%, Simmental breed and the crosses take around 1% and the other local breeds reach 15.5% of cow population. Black and White breed is mainly extended at the low land, coastal area and partly in higher regions. Jersey is extended throughout the whole country especially in the hilly and mountain areas, typical for the North part of the country.

The predominant sheep breeds are: local breeds account for 26.6%, *Tsygaia* breed and crosses 45.8%, Merinos sheep breed takes about 11.5%, Ruda flock – 9.5%, etc. *Tsygaia* breed flocks are located in the Southern and South-East hilly areas, as well as in central part of Albania. Local breeds are located in the North, North-East and partly in the centre of Albania. Farmers of the hilly and mountain areas prefer to manage the local sheep breeds because of their very good resistance to the diseases and moderate food requirements. Especially *Tsygaia* sheep crosses are preferable.

Local breeds account for 99% of goat population and the cultivated breeds 1% only.

In Albania only cultivated pig breeds are used in pig production.

Poultry production under extensive management use local breeds, while industrial sector intensive production is using cultivated breeds of high genetic production capacity.

Nonius, Haflinger and Arab equine breeds are located in the low land, whereas the local breeds, which take about 58% of equine breeds in the country, are located at the hilly and mountain areas. The farmers use them mostly for heavy works and transport.

Donkeys are completely of local breeds.

Livestock production systems

Livestock production has not the characteristics of a sustainable production system. It was mainly kept in very small, fragmented and mixed private farms, where a traditional production system is used. The average land size per farm used to be 1.4 ha.

The main livestock production systems are:
- *Production system of low input*. In this system mainly small ruminants are involved:
  o 95% of sheep
  o 99% of goats
  o 88% of cattle
  o 70% of pigs and
  o 50% of poultry
- *Production system of medium input:*
  o 12% of cattle
  o 30% of pigs
  o 20% of poultry and
  o 5% of sheep
  o 1% of goats
- *Livestock production system* of *high input* as the third one involves about:
  o 0.1% of dairy cattle
  o 30% of poultry

In 2003, 1060277 tons milk was produced, out of which 85% was produced by cows, 7% by sheep and 8% by goats. 123482 tons meat was produced. Farms for cattle fattening do not exceed 10 - 15 heads. The recent tendency is using crosses between dairy and beef breeds to increase beef production. These farms operate by their own input, exploiting natural resources mainly and land they have to their own disposal. Some of the fattening cattle farms represent the *medium input* production system. These farms' production level is more consolidated.

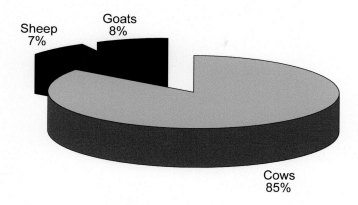

Figure 1. Milk production in year 2003.

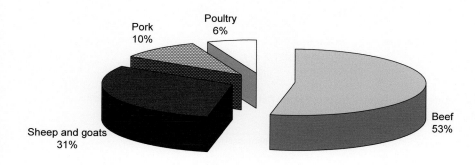

Figure 2. Meat production in year 2003.

Livestock farm size, management and extension needs

In Albania, livestock farms are composed of various species of animals, like cattle, small ruminants and poultry, while the number of pig, cattle or sheep and goat farms is low. The period after 1998 was characterized by a very intense organisation of new livestock farms. Livestock farms are totally privatised and do operate separately. Specialised livestock farms started to become more evident after 2001 (Table 1).

Table 1. The structure of farms with cows, sheep and goats, pigs and poultry.

Farms with cows	Cow per farm	No. of farms
	1 - 5 heads	318.500
	6 - 10 heads	540
	11 - 50 heads	125
	> 50 heads	9
Farms with sheep and goats	Sheep and goats per farm	No. of farms
	100 – 200 heads	69
	> 200 heads	30
Farms with sheep	Sheep per farm	No. of farms
	100 – 200 heads	1,186
	> 200 heads	534
Farms with goats	Goats per farm	No. of farms
	100 – 200 heads	683
	> 200 heads	162
Farms with pigs	Pigs per farm	No. of farms
	6 – 10 heads	140
	11 – 50 heads	100
	> 50 heads	6
Egg production centres	Egg production per farm	No. of centres
	10 – 50 thousand	11
	> 50 thousand	5
Meat production (Broiler)	Broiler production per farm	No. of farms
	20 – 50 thousand	10
	> 50 thousand	7

Source: Technical Report 2003 - Animal Production Department- MAF

These private family farms are usually managed as family businesses, using family land, housing area and family labour. The production technology is traditional.

The following farms are distinguished based on their private business, production systems, technologies, relation to the market and the economical potential:

- *Small farms*, which are highly fragmented, with the size of cow herds and sheep/goats flocks ranging from one to two cows and ten to fifteen sheep and goats. These farms are mainly located in the mountainous regions and hilly areas on the North - East of the country. Here, land size per farm does not exceed 1 ha. The farmers are not interested to cooperate with extension experts and do not realise the importance of all kind of information.

Small farms producing livestock products just for home consumption and to sustain country living, produce in a very extensive way. Farmers have insufficient knowledge in livestock breeding techniques and the farm animals are of inappropriate breeds. The lack of training leads to misapplication of breeding techniques, inability to produce quantity and quality of products for domestic markets.

- *Medium farms,* producing livestock products not only for home consumption but for the domestic market too. These farms are mainly located on low land and hilly areas. They own and manage not only local breeds but also pure-breed animals and crosses. Therefore these farms have good yields. But they have not a very strong or sustained linkage with the market.

  Production is performed in a traditional way but the farmers tried to have a good farm management and product quality control, to apply new breeding techniques or improved production technologies (e.g. using modern milking machines). The farmers possess cooling tanks and are entering into regular contracts with the collection units or dairy plants.

  The farmers want to be well-informed and well-trained in respect to the way of treatment and management of the genetic resources they have in their ownership. They participate in demonstration days and are aware of the need of the assistance of extension experts (*public extension service is run, financed and technically supported by the Ministry of Agriculture and Food - MAF*). The extension service network operates through out the country. This service provides advice, organizes practical seminars to the farmers, disseminates leaflets, which deal with the most crucial problems the farmers face (e.g. - application of artificial insemination, tracing and prevention of diseases, construction of the appropriate stables, feeding, etc.).

  Farms of medium level are often not able to face changed market conditions and to reach quality standards of products. The lack of capital and market demands (especially butchers and consumers who prefer to consume baby-beef) are forcing farmers to sell young animals (at two or three months) or at low slaughter weight (90 - 150 kg), thus reducing their profit. The lack of capital blocks the investments for further development of livestock production.

  Farmers are not cooperating with each other. This blocks possibilities of being well organized in associations, which could provide easier access to credits or markets to trade the products.

  In a number of private farms some breeding programs and projects are implemented to provide sustainable livestock development or its performance - farm applied research under the responsibility of MAF the Animal Husbandry Institute.

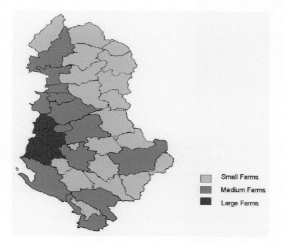

Small Farms
Medium Farms
Large Farms

Figure 3. Geographical distribution of small, medium and large farms in Albania.

- *Large farms* are mainly located in the low land areas. There are nine private cattle farms, which breed up to 100 heads of dairy cows. Poultry industrial establishments are characterized by contemporary or modern technology and high production capacity. So, in a short time the needs of Albanian people will be totally fulfilled with eggs. There are more than 30 private family farms, which are breeding mixed flocks of more than 300 sheep and goats. These farms are characterized by very strong and sustained relations to the domestic market.

On private farms or industrial centres modern production technologies are being applied right now and new elements of farm management are used, which actually improve the production, reproduction performances and feeding efficiency according to the breeding programs.

Recently, the Herd Book of Jersey and *Tarantaise* cattle breeds has been opened, which registers about 2000 head of cows.

The production system is limited by the following factors:
- application of traditional methods which cause weakness of farm management;
- lack of capital for applying or implementing new technologies;
- lack of organized infrastructure for collecting and processing of livestock products and sustainability of markets and prices;
- dairy industry is not able to absorb all milk produced or to process it into dairy products;
- Another limiting factor of farmers in the hilly and mountainous areas remains the long distance from the trade markets. They work seasonally exploiting their own input, natural sources and their land.

Figure 4. Picture of a modest farm in Albania.

Supporting policies for further sustainable livestock development:
- We have already begun to compile and adapt the legislation frame under the *acquis communitare*. So, the "Livestock Breeding" law is being prepared, which will fulfil the needs and deficiencies observed and expressed in "The Animal Origin Genetic Resources Situation in Albania" report.
- During 2003, Albania joined the ICAR organisation, which means not only the assistance but to accomplish the standards the association applies to the animal production control.
- What we are doing right now is modernization and improvement of milk production, as there is no milk quota in Albania.

- A priority is the protection of the local (authentic) breeds, which exist in considerable numbers in Albania. Protective policies are directed towards promotion and encouragement of the use of the products of these local breeds.

Figure 5. An impression of the country side of Albania.

Prospects of quota and farm management in Turkey

Çağla Yüksel Kaya[1] & Numan Akman[2]

[1]*Cattle Breeders' Association of Turkey, Konur 2 Sok. 71/6, Ankara, Turkey*
[2]*Ankara University, Faculty of Agriculture, Department of Animal Science, Ankara, Turkey*

Structural changes in near future

According to the provisional results of the agricultural census in 2001, there are about 3 Mio agricultural holdings in Turkey. Within these, mixed cropping-livestock holdings constitute the greatest share of all farm types with about 2 Mio holdings (67.42%). The number of holdings of specialist livestock is more than 72 thousand, which has a share of 2.36%, in total. The specialist livestock holdings are mostly found at the south-eastern parts of Turkey where the size of herds are relatively small and most of the farms don't have land. Furthermore, these farms hold 3.45% of the total cattle population and 9.19% of the small ruminant population. The change in the number of all agricultural holdings, mixed cropping-livestock holdings and the specialist livestock holdings between 1991 and 2001, is given in Table 1.

Table 1. Number of holdings in 1991 and 2001[15].

	1991	2001[16]	% Difference
Agricultural Holdings	4,068,432	3,075,515	-24.40
Mixed Cropping-Livestock Holdings	2,935,055	2,073,600	-29.35
Specialist Grazing Livestock Holdings	139,692	72,429	-48.15

Source: The State Institute of Statistics

There has been an obvious decrease in the number of all types of agricultural holdings. The decline in the number of both the agricultural and the cattle holdings shall be expected in the near future. The rate of decrease depends on the changes, which will occur in the agricultural sector as well as the other sectors (Akman *et al.*, 2000).

The cattle breeding include mostly the dairy cattle husbandry since the specialised beef cattle production is not very common. Most of the meat is obtained from the dairy exotic breeds, crossbreds, and local breeds. The change in the number of milking and total cattle by years is given at Figure 1. When the Figure 1 is examined the slow decline in the total number of cattle since 1999 could be seen.

---

[15] The numbers include the agricultural holdings found at all of the villages and the districts where the population is less than 5000.
[16] The results are provisional for 2001. According to the provisional results of agricultural census of 2001, there are 4,106,983 house holds dealing with agriculture found at all of the villages and the provinces and districts where the population is less than 25,000.

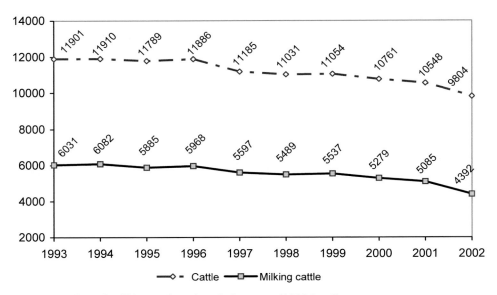

Figure 1. Number of milking and total cattle by years (1000 head).

Increase in the number of dairy cattle and total cattle population is expected in the long term. This increase could be grouped in two parts. One is the increase in number of cattle at the regions with a concentrated population and a relatively high income, where it is possible to grow forage crops with respect to climate, irrigation, etc. The other is the increase at the regions where forage crop growing and milk market opportunities are relatively lower. Farms that would be established at regions with appropriate marketing and production conditions would work with breeding cattle, which has high milk yield, produce towards the demands of the market, use the production techniques based on knowledge and technology. Furthermore, they are estimated to be middle- and big-scale holdings. These holdings would begin to be established at Marmara and Aegean regions and then would spread to the south-eastern parts of Turkey. The second type of holdings would be middle-scale holdings that provide the meat sector with cattle and aim to produce milk with relatively low capital. These holdings would be concentrated more at the east of Turkey. The number of holdings of private enterprises with big herds is predicted to be limited (Akman *et al.*, 2000).

The current statistics doesn't give detailed information about the farm incomes because there isn't a farm accountancy system in Turkey.

The farms have mostly mixed production because farmers consider livestock as cash. Whenever they need money urgently, they sell their animals. Thus, livestock becomes a kind of assurance for the farmers. In the next five years, it is assumed that the number of specialised farms is going to increase with the stability in milk prices.

The herd size is relatively very small in Turkish dairy herds when compared with the EU average. The average herd size of Turkey in general is 5 heads of cattle. Among all the cattle farms, 84% have less than 5 heads (Anonymous, 2001a). The herd size average increases to 26 cattle per farm at the herdbook system. Even if the farms don't specialise, the herd sizes will increase when the cattle number increases as the number of agricultural holdings decrease. The prediction of the increase in number of specialised farms could also lead with bigger cattle herds.

The average milk production of cow is another topic, which is rather low when compared with EU. The average milk yield per cow in Turkey changes between 1,800 and 2,000 kg (Akman *et al.,* 2000). However the average milk yield of the recorded cattle to the herdbook system is much higher. The breeds of the recorded cattle cause the big difference in the yield, which mostly covers purebred Holstein-Friesians, while the crossbreds of various exotic breeds and the local breeds (36,58%) constitute the general cattle population of Turkey (44,45%). The distribution of the breeds in the herdbook system is given in Figure 2. The trend of the average milk yield in herdbook system by years per breed is given in Figure 3.

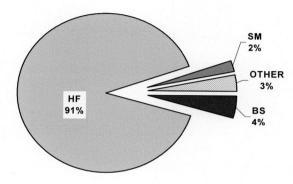

Source: Cattle Breeders' Association of Turkey

Figure 2. The distribution of the breeds in the herdbook system (26.08.2004).

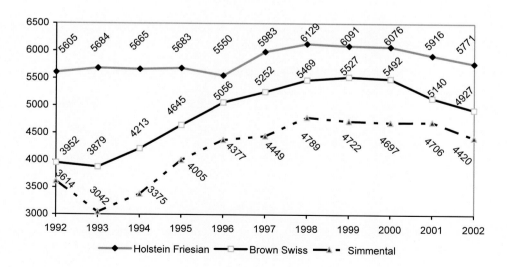

Figure 3. Average milk yield in herdbook system by years per breed (kg).

The ongoing decrease of average milk yield of recorded cows since 1999 is due to the increasing membership of the new eastern provinces and the state farms to Cattle Breeders' Association of Turkey (CBAT).

For the near future, the trend of average milk yield per recorded cow could be predicted as an increase since the main three breeds show a clear increase in the past ten years.

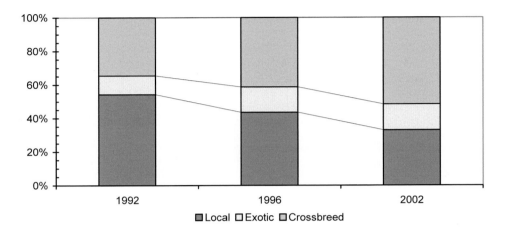

Figure 4. Distribution of breeds in cattle population (%).

The ratio of the local breeds, crossbreds and exotic breeds are changing sharply. According to the trend, it could be estimated that the ratio of crossbreeds would increase as the local breeds decrease and the rate of exotic breeds will go on increasing slightly over the next five years.

According to a projection study, the estimation of average milk yield of these genotypes for the next five years is given in Table 2.

Table 2. The estimation of average milk yield of genotypes (kg)[17] (Akman *et al.*, 2000).

Years	Exotic	Crossbreds	Local
2004	3,604	2,406	800
2005	3,676	2,442	800
2006	3,768	2,490	800
2007	3,862	2,540	800
2008	3,958	2,591	800
2009	4,057	2,643	800
2010	4,159	2,696	800

As given in the table, the average milk yield of local breeds are assumed as stable when the exotic breeds and their crossbreds have an increasing average milk yield due to the genetic

---

[17] It is estimated that average milk yield would increase 2.5% at exotic breeds and 1.5% at their crossbreeds per year.

improvement studies together with the progressing extension services, the increasing usage of new production techniques and farm management systems.

Although there are not any statistical information concerning the dairy farm workers, livestock farming in Turkey is generally a small-scale family farm activity, carried out in the vast majority of cases as a sideline activity by arable farmers (Anonymous, 2003a). There are studies of Dernek (1997) and Kıral (1997) for the calculation of demand and supply of agricultural labour but the assumptions don't have the data for labourer number per farm. Therefore, it is not possible to make a projection for the near future, as the current situation isn't known.

The total number of dairy factories is approximately 4000, in which 1300 have 1000 tonnes/year or more processing capacity. Only 10% of the current enterprises could be defined as modern processors. The cooperatives have a share of 6% in this 1300 dairy factory. Only 19.3% of the milk produced was sold to the dairy industry in 1996 (Anonymous, 2001b). Most of the milk produced (60 – 70%) is sold by the street sellers, the producers directly or at the small stores as fresh liquid milk, butter or yoghurt. Cheese and yoghurt are produced mainly at the specialised dairies. The number of dairy factories is expected to decrease in the coming years. It would be very hard for the small dairies to compete with the big dairy factories especially when the consumers are more informed. Moreover, it is harder for the very small dairies to get a production license from the Ministry when they don't meet the required sanitary rules. Besides, like the concentration in the dairy farms, the dairy factories shall have the same trend effect by the developments in the market. There would be less factories with higher processing capacities in the coming years.

Agricultural production differs from one region to another at the current situation. The eastern part of Turkey is still using the primitive technics of production while the western part makes use of the newest technological developments. A very complicated and detailed project is being implemented at the southeastern Anatolia to decrease the regional differences. However, the western parts are nearer to the big metropolis, which enables them to sell their products easier and with relatively high prices. The regional agricultural differences will continue as long as the rural economy, in general, is not improved and the immigration continues. Besides, there are big differences in the ecological situation throughout the country. It is considered that there are nine agricultural regions in Turkey and the agricultural analysis are carried out according to these regions. Each has its own characteristics concerning the climate, flora, and culture. If the planing of the projects is prepared considering these natural conditions of the regions, regional differences could be decreased and in the meantime, regional development could be provided.

The specialisation of agricultural holdings depends on the organisation of the agricultural markets and the regulation of the prices. The markets are not well organised at the current situation and the prices can not be foreseen because it is mostly determined by the organised industry. The dairy farmers' organisation is still not powerful enough to make an influence at the milk prices. Furthermore, the dairy policies do not have an intervention system for the milk and milk products which causes the farmers to sell all their production at whatever the price is given by the industry. Briefly, the dairy policies of the governments do not provide a fair income to the farmers and maintain the sustainability of the sector, which effects the agricultural structure. Most of the farmers prefer to have a mixed farming to secure themselves of the big changes occurring at the markets.

The mixed farming shall also increase for the dairy farms not to be obliged to the roughage, which would increase the production costs when bought. Moreover, it would be economic for the farms to continue on mixed farming when the farm sizes and the land structure are taken into account. The farm sizes are rather small and the farmlands are parcelled in many and small pieces. Thus, the dairy farms shall grow forage crops to decrease the production costs, even if they are specialised and big herds.

The rural development studies have started long ago in Turkey but when examined it is seen that the projects implemented were planned centrally. The centralised planning can not be successful unless it really refers to the primary problems of the region. Less achieved progress when most were a waste of time and money. These unplanned policies resulted in rural exodus with diminution of the rural population day by day. The rural industry, in this context, couldn't be developed enough. So most of the farms are full-time farming. Nevertheless, the agricultural workers usually work part-time, working at the field cropping and permanent cropping holdings in particular seasons and they are mostly migrant families. These workers have little connection with animal production and their numbers are decreasing every year depending on the increase in the mechanised farming. Moreover, the workers of the cattle farms are usually full-time worker families staying at the farms.

The part-time farming will lessen for the farmers themselves and perhaps increase, if the policies won't change, for the agricultural workers. Another important issue for Turkish agriculture is that it is still characterised by hidden unemployment, which is an enormous challenge in the economic development (Anonymous, 2003a). Hidden unemployment could be considered as an indication for the existence of a potential for the increase of part-time farming because the hidden unemployment could revert to part-time farming in the meantime.

The current policies encourage the new farmers to construct their farms more technologically and to continue their production more intensive. The herds are getting bigger and intensive as the producer prices proceed more stable. Moreover, the pasture quality is following a decreasing trend because of the incorrect use until now. Although new legislations are put into force to provide the improvement of the pastures, especially the western parts of Turkey will show more intensification of the farms. Another major reason for the intensification is the organisation of the farms and being able to exchange ideas and new technologies between themselves and also with the extension services of these organisations. The farmers of the eastern parts are not as well educated as the western parts, which doesn't enable them to organise and make use of the power of the civil societies. Therefore, intensification of the farms could be expected in near future as the herd sizes increase and the quality of milk becomes more important for the industry in particular at pricing.

The producer prices for milk have increased steadily since 2001. The increase between 2000 and 2002 has been less than expected but when we consider that there had been an economic crisis in Turkey in 2001, we could explain the decrease in the milk : feed ratio. When we look at the prices for 2004, calculated in June, we see that the milk : feed ratio started to fall down after its steady increase in 2003. The producer prices for milk and feed prices are compared at the following graph.

The milk prices show an increasing trend when it is examined monthly. This is a result of the dairy policies of the government as well as the strengthening social solidarity of producers. The subsidies given by the government to the specialised herdbook dairy farms and the encouragement of constructing better-equipped farms have increased the milk prices. If the national policies continue to support the system and organised farmers, the prices shall show an increasing trend. Although the prices have begun increasing, this won't be a stable trend unless either the government or the farmers themselves regulate the market.

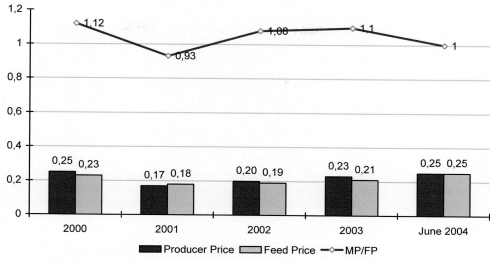

Source: CBAT

Figure 5. The change in the milk (MP) and feed (FP) prices between 2000 and 2004[18] (€/l).

National support

There are specific subsidies given to the dairy and beef farmers since 2000. The amounts for subsidies change every year. The subsidies given to the dairy farmers in 2004 are as following:

- *Milk Premium:* This premium is given to the farmers, for per litres of milk sold to the approved and licensed dairy factories. It is given 20,000 TL/l. (0.01 €/l), but when the farmer is registered to the herdbook system, the premium increases to 40,000 TL/l. (0.02 €/l). There are two aspects of this premium. First is to encourage the farmers to sell their milk to the licensed dairy factories and the second is, to promote the farmers to be involved to the herdbook system.
- *Artificial Insemination Premium:* This premium is given to the farmers who inseminated his heifers/cows artificially. The amount paid to the breeders' is 7,500,000 TL/AI/year/head (4.36 €/AI/year/head). If the breeder is registered to the herdbook system, or he lives in a province which has priority in rural development, then the premium increases to 15,000,000 TL/AI/year/head (8.72 €/AI/year/head). The aim of the premium is to stimulate the AI practices, the usage of good genetic material among the breeders and the registration to the herdbook system.
- *Calf Premium:* This premium is given to the calves born in the registered herds from registered cows to either the herdbook system or the pre-herdbook system by AI. The amount given to the pre-herdbook farms is 30,000,000 TL/head (17.44 €/head) and

---

[18] Euro – Turkish Lira conversions belong to the statistic year average. (2000: 573.942; 2001: 1.093.683; 2002: 1.429.766; 2003: 1.685.301; 2004: 1.720.115). Source: Central Bank of the Republic of Turkey

60,000,000 TL/head (34.88 €/head) for the herdbook farms. This subsidy is given to support the breeders who are registered to one of the recorded systems and to widespread the AI practices.

- *Pregnant Heifer Subsidy:* This subsidy is given to the purchasers when they buy a pregnant heifer with pedigree or certificate. Certificate is given to the heifers recorded mostly in the pre-herdbook system with only insemination data and pedigree is given to the heifers recorded only in the herdbook system with production data. When the purchaser buys a pregnant heifer with certificate, he gets 200,000,000 TL/head (116.27 €/head); when he buys a pregnant heifer with pedigree he gets 400,000,000 TL/head (232.54 €/head). The aim of this subsidy is to encourage the new farmers or the existing ones to purchase breeding cattle to their farms with certificates or pedigrees. Thus, with the effect of this subsidy the demand for the breeding cattle increases when compared with past and the specialised breeders gain more than the others do.
- *Subsidy to the Forage Crops:* This subsidy intends the growing of forage crops in the farmlands in order to decrease the costs of feed and support the animal husbandry. It differs between the single-year forage crops and the multi-year forage crops. The operational and investment expenses are covered by this subsidy. The amount differs from 20 – 35% of the total expenses.
- *Meat Premium:* This premium is given to meat of the cattle slaughtered at an approved and licensed slaughterhouse. The premium is 1,000,000 TL/kg (0.56 €/kg) and aims to prevent illegal slaughtering. Another aim is to support the national identification and registration system by giving the feedback information of slaughtered animals by the slaughterhouses to the authorities. However, this subsidy is given in different periods and years according to the changes in the meat market.
- *Support to the farms free of diseases in Thrace region:* Thrace region is located in the north-western part of Turkey, the part found in the European continent. This subsidy aims to increase the number of farms free of Brucellosis and Tuberculosis in this small region. With this purpose, the farms in this region fulfilling the health requirements are paid 50% more at milk premium. Moreover, they get 40,000,000 TL (23.25 €) per animal.
- *Low interest credits to dairy farms:* These credits are given to the farmers who constitute new farms with 10 or more than 10 heads breeding heifers, and to the existing ones who wants to enlarge the farm unto 10 heads of breeding heifers. Therefore; the rate of interests for operational and investment credits are decreased 40%. This credit encourages new farms as well as enlargement of the existing ones.

Farm management

The main problems in herd management are feeding, heat detection in cows, milking and milking hygiene. Most of the farms don't make rations and feed the animals according to their needs. Cows produce less than their potential because feeding isn't well organised at herds. There are economic losses due to the missing of the heating of the cows. The interval between calving and new pregnancy is long. Because of small-scaled herds, the milking is carried out manually to buckets and in order not to increase the production costs, most of the farmers don't use disinfectants before and after milking. Thus, in one hand mastitis is a very big problem at the herds while on the other hand milk doesn't meet the qualifications required by neither the industry nor the new regulations.

With the new milk hygiene legislation coming into force at January 2005 and the new support scheme for cattle breeding, it is expected that the milking systems and cooling of milk at holdings would develop. As the competition forces, milk industry would require qualified cooled milk. The demand of the market shall force the producers to produce with better technologies and techniques. The market demand would also lead to keeping records at farm level and the improvement of the current herdbook system. The feedback of the records shall also gain more importance for herd management than its actual situation. Thus, the other current problems would be overcome when the milk prices are stable. This would encourage the producers to improve their production systems. They would make more use of the recent developments in herd management.

There aren't sufficient data to give exact rates for the farm costs. However, as usual the biggest cost factors are feed and veterinary costs. Especially the share of feed in the total costs at an intensive cattle farm is as high as 60 – 70%. Inan (1992) has calculated the proportional distribution of the variable costs of farms in Tekirdağ in his study. The share of feed costs was found 79.76%. The following main cost factors of the farms were found the preparation of feed (5.94%) and milk given to calves (5.78%).

Organisation of extension and extension needs

The government, civil societies, and private sector give extension services. There are different departments of various ministries giving public extension services. The Ministry of Agriculture and the Rural Affairs (MARA) has a specific extension department and it gives service in all the provinces and districts via its Provincial and District Directorates. MARA has organised its provincial departments according to the Training and Visit System. The other public services use the methods appropriate to the region they give service and work. Not using the participatory oriented methods is what the public extension services have in common. The public extension programs are planned from top to bottom, which doesn't take into consideration the needs of the farmers (Tatlıdil *et al.*, 2000). There is also the new project of the MARA. They have appointed 1000 so-called agricultural volunteers to 1000 village. Although they are called volunteers, the local public administrations and/or relevant private sector pay them. They are in charge of informing those villages with the recent information and giving services. They could be either veterinarians or agricultural engineers. Since it is a very new project, the achievement can't be determined. Besides the extension departments of the ministries, there is the public agricultural TV channel, which is more useful to the farmers because audiovisual presentation is more effective. The agricultural channel broadcasts only half a day.

There are also some non-governmental organisations and charities, which give extension services. Especially the specialised organisations like leader farmer organisation, cattle breeders associations, rural development cooperatives, irrigation cooperatives…etc give extension services to their members in specific subjects which are usually planned according to the real needs of the farmers and achieve the intended purposes.

Third kind of extension services are given by the private companies like agricultural processing industry, semen sellers, milking equipment manufacturers, etc. They have more chance to inform the farmers face to face and they are more aware of what the farmers need knowledge on. Especially the agricultural industry informs the farmers on production systems to find raw material at the quality and quantity they need. Besides, the financial aspects of these training courses and various extension services are covered with the products bought or sold by the companies.

Although the extension services try to inform the farmers about the national and European Union's regulations and rules, the farmers who have knowledge about the subject and who are interested are not many in number. But as the communication technologies improve, more farmers will get to know the international developments easier and quicker.

The extension services have played a role in introducing new systems as well as traditional production techniques. Examples could be given for protecting the animals against the effects of hot weather, spreading the AI practices, separating the milk with antibiotic, cooling of the milk, cleaning of the teats, calf care…etc. Both the public and the private extension services have given information about management practices for years and they have achieved to reach the farmers in some issues.

At the current situation, the personnel carrying out the extension services aren't all specialised in extension. Turkey needs extension experts to carry out the extension services professionally. Besides, there aren't enough personnel to carry on the necessary extension services. Extension services have always been a very expensive tool for public and civil societies to give and this is another requirement of the Turkish informing system. The expenses not only include the experts but also the necessary equipment as well. The private extension services should be generalized but the current market conditions don't enable the farmers to pay for information. Furthermore, the information has been given for free to everybody by public authorities until now. This is another factor impeding the farmers to begin paying for information and extension services they get.

The organisation of the extension projects shall be planned towards the needs of the farmers, with the participation of the farmers and the most appropriate methods of extension shall be selected in order to be successful. Especially the farmers shall be correctly and fully informed about the EU legislation and practices. Unless the farmers support the accession to EU at field practices, the adaptation of the legislation will be insufficient for the participation.

The funds for extension services shall be increased in the institutes and more private extension companies shall be established in order to give better service to the farmers. The farmer organisations shall give more importance to the extension activities and farmers shall require better services from the public, private and institutional extension services. Specialised service shall be given by the institutions to achieve the goals of the projects.

The biggest challenge for the dairy industry in years ahead

The biggest challenge for the dairy sector shall be the international competition at the world markets. If the policymakers don't consider the current problems when making and implementing the new policies, the number of cattle could continue to decrease without an increase in milk yield. This will lead to importation of cattle and dairy products. The numbers show that Turkey has a deficit in animal proteins. When projected for the next ten years, Turkey shall implement a long-term dairy policy in order to overcome this deficit. It is expected that if policies will encourage dairy breeding and genetic improvement, in particular for milk proteins, the deficit could be overcome and the level of animal protein per capita could reach to the world level. Along with the augmentation of the cattle population and yield, the quality and hygiene should also be taken into account. As long as the milk produced at farms does not meet the international quality requirements, the industry won't be able to compete at the world markets.

Furthermore, a possible negative development in the cattle breeding sector shall also lead to a deficit in cattle meat. While this could result in the importation of cattle meat, it would also impede the exportation potential of particular regions of Turkey, both for meat and dairy products.

To summarize, long-term policies shall take care of the farmers income, product quality and the enlargement of herd sizes to accomplish an economical production system while taking into consideration consumers and international competition as well.

Figure 6. Map of Turkey.

References

Akman N., K. Özkütük, S. Kumlu & S.M. Yener, 2000. Türkiye'de Sığır Yetiştiriciliği ve Sığır Yetiştiriciliğinin Geleceği, Türkiye Ziraat Mühendisliği 5. Teknik Kongresi, 17-21 January 2000, Ankara: 741-764.

Anonymous, 2001a. Sub-committee Report of Animal Husbandry, Animal Husbandry Commission, State Planning Organisation, Ankara: 134-162.

Anonymous, 2001b. Sub-committee Report of Milk and Milk Products Industry, Food Industry Commission, State Planning Organisation, Ankara

Anonymous, 2003a. Country Report Turkey, Agricultural Situation in the Candidate Countries, European Commission Directorate- General for Agriculture

Anonymous, 2003b. Animal Husbandry Report, Union of Turkish Chambers of Agriculture

Dernek, Z., 1998. Agricultural Labour Demand in Turkey, Agricultural Structure and Employment in Turkey, State Institute of Statistics Prime Ministry Republic of Turkey, Ankara: 129-187.

İnan, İ.H., 1992. Tekirdağ İli Süt Sığırcılığı İşletmelerinin Doğrusal Planlama Yöntemi ile Planlanması ve Planlı Çalışmanın İşletme Gelirine Etkisi, Trakya Bölgesi 1. Hayvancılık Sempozyumu, 8-9 January 1992, Tekirdağ: 261-275.

Tatlıdil, H. & C. Ceylan, 2000. Türkiye'de Tarımsal Yayım Hizmetlerinin Geliştirilmesi, Türkiye Ziraat Mühendisliği 5. Teknik Kongresi, 17-21 January 2000, Ankara: 1105-1115.

The state of livestock production in the Republic of Belarus

Ivan S. Kyssa

International Public Association of Animal Breeders "East-West", Kigevatava 7-153, 220024 Minsk, Belarus

The leading agricultural branch in the republic is livestock production, rationally combined with the cultivation of grain, sugar beet, potatoes and vegetables. The average number of workers in cattle-breeding from 1999 to 2001 was 194.14 thousand people or 37% of all agricultural workers. The share of gross production of this branch in the structure of gross agricultural output is more than 45% and more than 55% when including the public sector.

In the past eleven years stock-raising branch has been characterised by the following negative tendencies: reduction of livestock number and decrease of stock-raising production funds. Since 1990 cattle livestock decreased by 2,946 thousand heads or 41%, including 594 thousand or 24% of cows; 1,773 thousand or 34% of swine and 421 thousand or 82% of sheep at all types of farms. The highest rate of livestock reduction is in public sector where the number of cattle decreased by 44%, of swine – by 43%, and of sheep – by 98%. For comparison, livestock of cattle at family farms decreased by 18%, of swine – by 15%, of sheep – by 63%. The only exception is poultry, its reduction in private sector was higher (56%) than in the public one (28%).

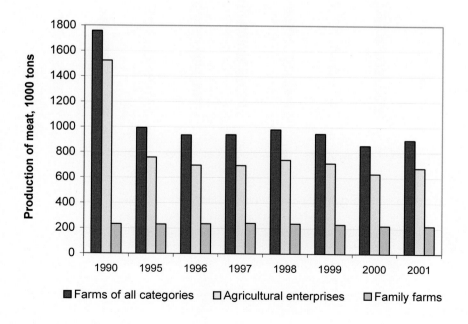

Figure 1. Dynamics of meat production (in live weight according to categories of farms).

In 2001 all types of farms produced 627 thousand tons of cattle and poultry (in slaughter weight) which makes 53% of the year 1990 level, 4834 thousand tons or 65% of milk, 3144 million or 86% of eggs, 163 tons or 17% of wool (in physical weight), 2420 tons or 60% of honey.

Milk production

Dairy cattle breeding as one of the main livestock production branches of agriculture in Belarus has got a relatively high economic development. In pre-crisis years (late eighties – early nineties) milk production in the republic reached about 7.5 million tons. It was enough to satisfy domestic and export needs of the country. In the eighties the branch was developing intensively.

In times of national economy destabilisation, including the agro-industrial complex, dairy cattle breeding which has taken 14 – 16% of resources spent for agriculture, and 28 – 30% of those spent for livestock production became unprofitable for most producers.

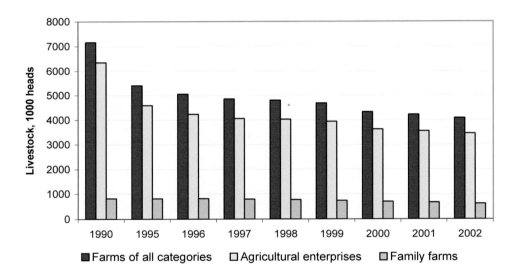

Figure 2. Dynamics of cattle population according to categories of farms.

Milk herds in the public sector of agriculture in the Republic of Belarus are present at three types of farms. *The first* type are dairy farms: they are oriented to milk production and calf-raising up to the age of 20 days. The size of such farms depends on specialisation level, fodder supply, environmental conditions, type of buildings and machinery system. In the Republic of Belarus collective farms, state managed farms and big farming associations house 200, 400, 800 and 1200 cows. *The second* type of farms are dairy-beef farms where both milk and meat are produced. Besides cow-sheds and calf-sheds there are buildings for young cattle production and cattle fattening. *The third* type of farms are beef-dairy farms where the main activity is the raising of young animals and fattening of cattle, while milk production is second branch.

Beef production

Cattle production is one of the main branches of agricultural specialization in the Republic of Belarus and as such represents the main constituent of the meat production sub-complex. The share of beef and veal in the republic is more than 40%. To develop this branch collective farms spend about 35% of the available funds and 45% of forage to this branch. The share of cattle production in agricultural commodity output makes 22 – 27%.

Actually, cattle breeding is the only activity which is allowed to use effectively grassland in large quantities of field crops, hay-lands and pastures to produce so valuable goods as milk and meat. In addition, cattle production gives raw materials for leather manufacture, and a number of different products (from soap to endocrine medications) made of the waste at slaughter-houses.

Cattle production in Belarus is represented mainly by dairy-beef livestock. In the nineties an attempt was made to breed specialised beef cattle in the zone of Belorussian Polesye, but on account of different reasons it did not develop as expected. During these years eight specialised farms (for pedigree cattle) were created, 5 thousand heads of young animals of different beef breeds (Charolais, Limousine, Light Aquitane, Men-Angeous) were imported.

The main beef producers in the diversified agricultural conditions were and remain enterprises of public sector. About 97% of fattening livestock and 92% of beef production is concentrated there. Cattle are kept in almost all agricultural enterprises and are relatively evenly spread over the territory of the republic.

In the private sector the curve of bred and fattened cattle went down sharply after the rise in the early nineties, and by the beginning of 1999 it reached the level of 67 thousand heads versus 415 thousand in 1992.

40% of family farms do not keep young cattle, 16% fatten one head, 26% fatten 2 to 3 heads and only 18% keep and fatten more than 3 heads. The share of beef of the total sales on family farms is 8% (the share of all stock-raising products in sale proceeds do not exceed 19%).

On the basis of the existing beef production technologies the following types of livestock production farms have been formed in the republic:

- Specialised farms engaged in the production and fattening of young animals, aged from 4 - 6 to 15 - 18 months, or those engaged in final fattening.
- Diversified farms with full production cycle where raising and fattening of young animals and their further sale is carried out by the same enterprise under the conditions of inter-farm specialisation. With the exception of specialised farms this group comprises practically all farms in the republic.

9 – 12% of cattle are kept at livestock production complexes of the republic and 10 – 12% of beef is produced there. Functioning of large enterprises over a long period of time depends on the expediency of their construction and their economic effectiveness.

Pork production

Pig breeding is of great importance as the most dynamic and prolific animal production branch. Its final product is meat and lard for human nutrition, as well as pigskin, bristle, etc. for light industry.

The need of pork to satisfy domestic and export demand is estimated at the rate of 400-450 thousand tons in slaughter weight, but the actual production as low as 1/3 in the recent years. At present the share of pork in the meat balance of the republic is more than 45%. It shows a growing tendency.

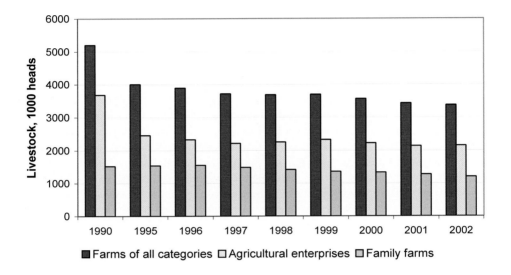

Figure 3. Dynamics of pig population according to categories of farms.

Pig production in Belarus is a traditional branch of agriculture with rather high level of development. In the nineties the number of pigs in all farm categories exceeded 5 million heads. Pig density per 100 hectares of agricultural land was more than 80 heads. From 1980 to 1990 pork production increased from 350 to 440 thousand tons. Profitability of production exceeded 35%.

However, from the year 1990 the process of decline has begun. Pig population in all categories of farms decreased by 1/4, and by more than 1/3 in the public sector. In the course of five years a sharp decline in animal productivity took place, and due to increase in labour costs and fodder inputs, production became unprofitable.

The situation has gradually improved starting from the mid-nineties. Pig production output per head started to grow and by the year 1998 reached 111 kg, offspring output per 100 sows approached 2,500 heads. Pork sales in all categories of farms reached 400 thousand tons, of which more than 40% was produced at family farms and farmer farms.

The most widespread pork breed in the republic is Large White, its share is more than 85%. Other breeds in the republic are Belarusian Black-variegated, Estonian Bacon breed and pigs of Belarusian meat type. More than 97% of pigs in Belarus are purebred.

Territorially, pig production is spread all over the republic. More than 60% of pigs are concentrated at the farms of public sector. Each administrative region has several agricultural enterprises engaged in pig production and fattening for the market.

The prevailing processes in the republic are concentration and specialisation. In Belarus about 110 large pig production complexes were created, meant for the production and fattening of 12 - 24 - 54 - 108 thousand heads a year. About 1.5 - 1.6 million pigs or 65 – 70% of total livestock population are kept and concentrated at agricultural enterprises where 170 - 180 thousand tons, or 75 – 80% of meat is produced.

Sheep production

Sheep production in Belarus has always been considered as a supplementary branch of animal production and was organized on extensive basis. In order to increase productivity of sheep breeds in the republic (Prekos, Romanov, Latvian Dark-head), to intensify production and to improve produce quality, specialization of sheep-breeding took place in the nineties. 125 specialized farms were created, the capacity of which was 3 thousand animals. At these specialized farms we developed and introduced progressive technology which allowed to increase gross wool production in the republic up to 9060 centners and mutton production up to 8 thousand tons. In the years 1984 - 1987 clipping of wool made 2.3 – 2.4 kg per sheep, lamb output was 89 - 91 head per 100 sheep.

In the nineties rapid reduction of sheep took place in the republic. Their total number at the beginning of the year 2002 was 83 thousand heads, including 6 thousand heads of the public sector. Gross wool production in 2001 fell to 31 tons, wool clipping decreased to 1.7 - 2 kg per sheep. Considerable growth of livestock in the near future is not expected.

Eggs and poultry production

The main producer of poultry products in the Republic of Belarus is the republican association "Belptytseprom" (RA "Belptytseprom") which produced 1,863 million eggs and 95.5 thousand tons of poultry in 2001, which makes 98% and 93% of all agricultural enterprises' produce, 59% and 86% of farms of all categories in the republic. Considerable number of poultry (33%) is kept by the citizens. It holds 39% of total eggs quantity.

Seven largest broiler production units with yearly output of more than 3,000 tons produce 90% of all broiler meat in the republic.

Egg production in Belarus reaches the level of developed countries. In 2001, yearly output was 341 eggs per capita. Presently, the highest figure of egg production in the world goes to China: 340 - 350 eggs per capita. Each citizen of Belarus consumed 226 eggs in 2001. For comparison, the first place in the world according to the level of egg consumption per capita belongs to Japan (345 eggs a year per capita), the second belongs to China (288 eggs).

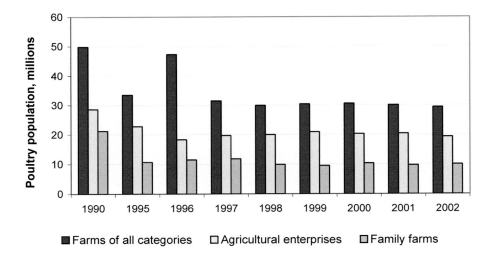

Figure 4. Dynamics of poultry population according to categories of farms.

Poultry consumption compared to eggs is very low in Belarus and makes 7 kg per capita. Its share in the structure of all types of meat consumption is 11%. It was calculated by the Belarusian Scientific and Research Institute of Agrarian Economy that production at full capacity according to the norms, and with average daily weight increase of 30 g per head, which is highly possible having sufficient fodder supply, poultry production in the Republic of Belarus could have been 15 kg per capita. Egg production could be increased by 25%.

Figure 5. Map of Belarus.

Developments of cattle husbandry in Georgia

Giorgi Saghirashvili, Tamar Kartvelishvili, Nino Kishmareishvili and Ekatherine Tsurtsumia

Georgian National Association for Animal Production (GNAAP), 4ª Marshal Gelovani str., 0159, Tbilisi, Georgia

Introduction

Georgia is located in the Caucasus on the crossroad of Asia and Europe on the northern periphery of subtropical zone. Georgia represents an eastern gateway to Europe along with the other states of the South Caucasus-Azerbaijan and Armenia.

Figure 1. Map of Georgia.

The geographical location of Georgia conditions the diversity of its nature. Almost all climate zones existed on earth are spread on the small territory of the country – from the humid subtropical climate to the glaciers and all types of soil. Abundant water resources of Georgia are used in irrigation, in the energy generating sector and for common water supply. Georgia has a great prospective potential to use its rich natural complexes for different purposes, for instance, for agriculture.

Specific natural and economic conditions greatly affect Georgian agriculture. In terms of modern borders, total area of the country territory is 69.7 thousand sq. km. with nearly 2.6 million hectares of agricultural land. The average use of land per person is 0.16 hectares of plough-land and 0.62 hectares of grazing land. Georgia is a country of high agricultural potential, but currently is not capable to use its potential completely. About 1.3 million people are employed in this sector, which is 55 of the total employment of the country.

In 2002, Georgia took 120th place in the list of 227 countries according to the GNP per person, representing the group of countries with low income in accordance with the international standards. Consumption expenditures of the household to farm ratio was 56%.

Agriculture is of vital importance to the Georgian economy, and its share in GDP accounts for 21%. This indicator reduced from 30% down to 21% during 6 years. This is caused by the poor agricultural infrastructure of Georgia, low level of agricultural production intensification and lack of modern scientific support and practical achievements in the production. There are serious problems in utilization of land resources: the structure of planted areas of food crops was destroyed. In fact, the material-technical base and agricultural equipment need to be re-established. Highly qualified specialists are not being trained complying with European standards. Investment activities also face huge barriers and the management and marketing systems are not effective.

The production level of animal food products does not meet the population requirements. The animal food product supply farms a big problem. Moreover, there is a high risk of pollution of these products, which may cause a serious danger for the human health. Food safety and quality do not comply with appropriate standards. To ensure safety of human health it is necessary to provide strict controls of the safety and quality of animal food products.

The structure of agricultural land ownership underwent a significant transformation in the first stage of the agrarian reforms. After the land reforms, about one million households became the owners of nearly 30% of total agricultural land.

From 1990 on the transition from the centrally planned economy to the market economy caused a crisis that, on its part, resulted in a reduction of the number of farm animals, decrease in productivity and a down fall in the animal production industry.

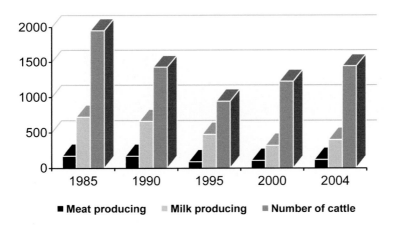

Figure 2. Structural changes in cattle production in years 1985 – 2004.

Population living in poverty, started to gradually come back to their villages and do small farming. Annual and perennial cultures compose the field structure of Georgian agriculture and livestock.

Nowadays, small (3-15 cattle or mixed types) and middle size (15 - 50 cattle or mixed types) livestock farms compose the main part of the household agriculture.

There exist large farms (50 - 250 cattle) mainly with dairy cows which have arable land for providing the feed base for the animals. There are plenty of part-time farms. A few farmers keep intensive breeds, such as Holstein-Friesians with annual average yield of 5,000 – 6,000 kg. These farmers produce and distribute locally various milk products for the domestic market. There exist 3 - 4 wide-scale milk factories which also produce more than 10 kinds of milk products for the domestic market. The price of milk per litre fluctuates according to regions and seasons from 15 to 50 cent (US). Since 2000 up to present the milk price rose insignificantly. However, the milk quantity is presently rising extensively without exception.

Formation of market related structures in the agriculture sector of Georgia, creation of new types of enterprises, redistribution of agricultural output in the private sector, further implementation of a consumer market under free trade conditions - require carrying out necessary reforms.

On farm level, the management of farms is problematic in Georgia. It's necessary to carry out trainings for implementing new affective management practices. The instruments for stimulating good management practices and the subsidies for supporting dairy, beef, suckler cows and sheep farmers are very poor.

Primary animal production systems in Georgia

There are three major production systems in Georgia which significantly contribute to food production and agriculture, rural communities and ecology: A low-input system – mostly non-certified organic production (ecological production), a medium- input system – mixed production.

1. Low-input system

The low-input system mostly refers to non-certified organic production. Non-certified organic production is substantially present at the small and middle family farms in all regions of Georgia. Certified organic production is a system of farming, which is today still relatively marginal in Georgian agricultural, but has a potential to expand due to natural conditions and a traditional relationship towards conservation of agricultural resources.

2. Medium-Input System

A medium-input system refers to a system of sustainable or basic production which is of mixed type-dependable on geographic, social and economic factors. The family farms, which belong to the medium-input system, are occupied in agriculture already for many years. Within the framework of a sustainable agriculture family farm, usually more than one livestock species is raised (cattle, pigs, poultry, sheep, and goats). As well imported as indigenous animal breeds are present in Georgia, but the genetic and phenotypic appearance is very poor.

Organizational characteristics of production systems

After the reorganization processes of agricultural enterprises the former collective farms, state farms and other types of agricultural enterprises, being founded on the private co-operative and other kind of ownership, were transformed into new legal-organization forms. Nowadays, the different forms of property act in the rural area are: companies of limited liability, companies of joint liability, joint-stock companies, co-operative farms and individual farms.

The sustainable agriculture system consists mainly of family farms, which implies private ownership. Most of them are of mixed production. Thus today the household is the main producer of output. However, they are oriented at self-provision and have a low level of goods output.

Figure 3. Stacks of hay – typical way of making hay.

Input dependence

The majority of family farms utilizes their own crop production for the provision of basic animal feeds and consequently do not depend on input from outside. They depend on veterinary care, genetic selection and counselling services and some of them also on the purchase of highly concentrated protein components, mineral and vitamin additives. This dependence is only partly filled in and that affects the productivity and health of the animal population.

Most important animal products

The following species are utilized in primary livestock production: cattle, pigs, sheep, goats, poultry, fish and bees. Cow milk is predominant in milk production, while pork and poultry are equally represented in meat production. Locally adapted breeds fulfil a much larger role in all livestock sectors than modern imported breeds. The reason is that high production breeds are simply not imported. Nevertheless, the productivity of the local breeds is low, because the breeds are degenerated.

The most important primary livestock products are meat, milk, eggs, fish and poultry. Georgian regions differ in respect of significance of these products. The importance of secondary products relates to particular regions, depending on geographic, social and economic status and management of natural resources.

In last years there has been a significant reduction in livestock products export, since we are not self-sufficient in livestock production. Social difficulties and privatization processes have substantially contributed to the decline in livestock production.

Table 1. Productivity.

	2000	2001	2002	2003
Milk-cow's average yield, kg	935	1,018	1,041	1,038
of which by agricultural enterprises, like co-operatives	873	786	915	835
by family farms, households	963	1,047	1053	1,050
Average yield of wool, kg	3.0	2.9	2.8	2.8
of which by agricultural enterprises, like co-operatives	1.2	1.2	1.3	1.2
by family farms, households	3.2	2.7	2.7	2.7
Average eggs production by all categories of farms, units	130	127	129	130

Table 2. Livestock production in years 2000-2003.

Years	Meat (in slaughter weight), tons (x 1000)	Milk in tons (x 1000)	Eggs, mln.units	Wool, tons
2000	107.9	618.9	361.4	1,860
2001	102.4	710.0	395.4	1,898
2002	106.9	742.1	408.8	1,994
2003	108.9	765.1	458.1	2,023

Table 3. Total milk production in years 2002-2003.

Years	Total milk, tons	In which	
		Dairy cows, tons	Sheep and Goats, tons
2002	742,120	720,703	21,417
2003	765,100	743,270	21,830

Major trends and significant changes in the use and management of animals

In the last ten years there have been significant changes in the use and management of animals. The changes result from the changes in ownership caused by the application of the Agricultural land Act. Objectives of the politics of changes in ownership structure are: an increase in the size of family farms, i.e. property enlargement through privatization, sale or lease of agricultural land that was before in state ownership. In livestock production it may lead to an increase in livestock funds and modernization in production capacities in order to improve product quality and provide population with safe food as much as possible by competitive domestic agricultural production.

The future courses of action for agricultural politics with the aim to restructure the agricultural sector are:
• try to achieve vital commercial farms;
• modernization of production capacities and agro-technological and agro-economic procedures;
• increase in the role of farmers on the agricultural product market;

- impact on the changes in the agrarian structure and production technology in order to increase competitiveness of agricultural production;
- implementation of financial resources and compensation in agriculture;
- reform of the funds for financing and subsidizing farmers;
- increase in efficiency of administrative and special services and associations in agriculture;
- stimulation of personnel training;
- implementation of measures for protection of biological and landscape diversity in farming;
- stimulation and development of ecological and traditional agriculture that allows the survival of a relatively rich animal world.

All this will considerably affect production systems in livestock production.

Today the major limiting factors and constraints which affect the productivity and efficiency in livestock breeding are small scale farms, still unfinished privatization, inefficient production levels due to insufficient education of the farmers, market instability, lack of processing and final processing capacities, lack of cheap capital and insufficient financial support. This combination of limiting factors is common for the agriculture in countries in transition.

Nowadays, the situation with the Georgian livestock and local breeds is disastrous. The extinction of endemic breeds will seriously damage the Gene pool of the country, because these breeds have characteristics exclusively to them and have no analogues anywhere. These characteristics are: high endurance to diseases (absolute resistance to leucosis and pyroplazmosis); adaptability to the temperature fluctuations and low oxygen consistence in mountainous areas; adaptation to steep pastures (30^0 - 35^0), which are practically impossible to be grazed by other breeds; specific taste of milk and milk products. These characteristics of the local Gene pool are achieved by selection carried out for a long time, representing a source for genetic completeness. AnGR are still preserved in small numbers. It is necessary to protect endangered animals in the cattle breeding field as a whole. However, during the past fifteen years, nothing was done to preserve and maintain livestock in the country. From a structural point of view, the breeding activity as such does not any longer exist in the country. The technical and material know-how and the livestock population itself are mostly destroyed. The professional level of the existing breeding specialists does not comply with modern requirements. The feeding base is poor and the artificial insemination network is destroyed. The insemination service of animals carries a chaotic character lacking any control and involving bulls of unknown origin, of low productivity and not free of diseases. Currently the farmers sector in Georgia suffers a crisis. Only a few farmers, who acquired the qualifications in Europe, meet reasonable standards of work.

There are few local breeds preserved in Georgian: Caucasian nut brown cow, Georgian mountain cow, Megruli red cow, Imereti sheep, Tusheti sheep, Georgian mild wool and half-mild wool fat tail sheep, Kakheti hog, Svaneti hog, Straw-colored and Megruli (Megrula) chicken, Georgian turkey, Javakheti goose, Unique Georgian bee, Tusheti and Megruli horse.

Caucasian nut brown cow is degenerated and endangered. During the period of collective farming in Georgia, Caucasian nut-brown sort composed 90% of the total amount, which was economically justified. For example, there were 1.1 million heads of cattle of the Caucasian nut-brown sort in 1990, of which 330 thousand heads of dairy cows.

By now, the number of cattle close to the Caucasian nut brown is more than 95% of the total amount of cattle, but their productivity level does not comply with the standards for modern breeds. At average feeding conditions, the volume of the annual production per cow is 2,400 – 2,800 kg with 3.8 – 4% fat. Under conditions of better feeding and maintenance, the milk level reaches 3,500 – 4,500 kg.

The record dairy production was 8,789 kg which demonstrates the high genetic potential in this breed. This is the breed in the country, of which the milk is used for making the Swiss cheese on the Alpine pastures.

Figure 4. Caucasian nut brown cow.

Georgian Mountain Cow is an oldest breed, first of all, of milk direction. It is also used as beef cattle and draught force. Presently it is preserved on the Southern slopes of Caucasus mountain range and in mountainous Ajara. The extension zones of this breed are rich of rivers and brooks heads, we rarely meet plains. In the most zones of extension the bent of pastures reaches 30 - 35^0 and other cattle could not use it, except Georgian Mountain Cattle. The winter food of Georgian Mountain Cattle is very few, which is caused by the lack of arable lands and stern climate. In many mountain villages the arable lands compose 3 – 5% of the total land area, unused rocky slopes, mountain ranges, forests and natural pasture-arable lands occupy the great part of the territory.

Georgian Mountain Cow is very small, the height in wither of the cow is averagely 98-100cm. Its colour is various: black, red and straw-coloured, black-motley, red-motley. It is characterized by low milk yield in the conditions of primitive feeding, but in the case of improved feeding and care-keeping the milk yield increases up to on average 2000kg with 4.2% fatness. Milk is characterized by small diameter of fat bubbles, which indicates its dietary peculiarities. During the increase of milk yield, Georgian Mountain Cattle maintains fat percentage composition in milk. This peculiarity distinguishes it from other breeds. It has a hard constitution, endurance, milk butter-fat and high culinary peculiarities of meat.

Megruli Red Cow represents the breed of universal usage. It is raised with the completion of local small-body cattle in West Georgia in 60-s of XIX century.

Megruli Red cow spent summer in alpine zones of mountains, but in winter it pastured in Kolkheti marshes without stationary and supplementary food. In nomadic conditions the milk yield of cows was increasing from 2-3 to 7-10 litter.

Figure 5. Megruli Red cow.

This breed was permanently in the open air, so this factor conditioned its adaptability towards local conditions, health endurance, hard constitution and good working peculiarities. Megruli Red cow is characterized by peculiar exterior. It is bigger than aboriginal breeds of Georgia. The constitution of this breed is mostly like to milk direction herd; In addition, it is characterized by hard working peculiarities, endurance, strength and quick movement.

Georgian pig

Three main sorts of pigs are extended in Georgia: Kakhuri pig, Svanuri pig, Abkhazuri pig. *The Kakhuri pig* is one the oldest breeds from European origin, that was the result of European wild pig's domestication. It is not characterized by high productivity. Kakhuri pig is liked by its best quality of taste. *Svanuri pig* is extended in the high mountainous zone in west Georgia (1800-2000 meters from sea level). By immunogenetic learning it's clear, that they are alike to their wild ancestors.

Georgian sheep

In Georgia sheep breeding is historically a traditional branch of animal husbandry. In the further past (900-1100 years A.D.), the breeding of mild wool sheep was very famous in the countries of small Asia and in west Georgia – in Kolkheti, from where they were broadly extended in the Mediterranean Sea countries. In ancient Kolkheti the merinos sheep breeding was developed, which is proved by the famous legend about the Argonauts travelling in Kolkheti for Golden Fleece acquirement.

Imeruli sheep is the descendant of old Georgian sheep. It is small and has thin, elongated body, short cone-shaped tail, hard extremities and dry constitution. This breed is mostly white, but sometimes motley or coloured. The unique breed of Imeruli sheep is characterized by expensive biological-productive peculiarities. It is inseminated at the age of 5 - 6 months. This breed is very productive and represents the expensive genetic material for creation of sheep new types and breeds. It gives pure, transitive-fleece wool of extra class and meat with good taste peculiarities, which has not specific smell. It is shaved three times in a year.

Imeruli sheep grows quickly and is early-matured. The lambs, at the age of 6 months reach 75% of grown-up sheep, at the age of 4 months it is sexually matured, at the age of 11 - 13 months it gives 2 - 5 offspring and grows 2 of them without any help. It multiplies in any season of the year (it is polyestric). National selection was directed to receive 5 offspring and multi-productive generation during many centuries. This breed is characterized by the quick circulation of generation, because it is early, polyestric and multi-productive, for example, the descendants of one 3 year ewe often reach 10-12 livestock, but the record indicator is 76 in 3.5 years.

Natural habitat of Imeruli sheep is reducing due to lack of village pastures. This breed is now preserved in Imereti, Racha, Lechkhumi and Svaneti. Tushuri and other breed rams were brought in West Georgia in the last years, which worsened the breed peculiarities of Imeruli sheep. It is most significant to use its unique gene fund in industrial sheep-breeding.

Tushuri sheep is a half fat-tail sheep breed. It is raised in nomadic conditions in East Georgia due to national selection (XIII - XIV centuries) by crossing of old Georgian sheep to other rough wool breeds.

This breed is white, but some of them have black or brown spots on the face and extremities. The rams have well-developed spiral horns but the ewes sometimes have small horns. They are on the pastures during all seasons. Tushuri sheep is compact, has a hard constitution, it endures the drive on far distances and eats well on poor pastures. They go 250-500k from summer pastures to winter ones in very difficult and stern conditions. These nomadic conditions impacted greatly on forming a firm constitution of sheep.

Its meat and wool productivity increases distinctly in the case of improved feeding. This breed matures early, is lack-productive, have high quality meat and white, flexible, elastic and glittering wool, of which people knit high quality rugs. A tasty cheese is produced from its milk, which is successfully used not only in Georgia.

Figure 6. Georgian sheep.

Georgian half-mild fat-tail wool sheep. This breed is raised from crossing of Tushuri sheep to Mild wool Rambuli and Prekosi rams. Then these crossed individuals had been selected and crossed to each other. Due to purposeful selection, new wool-meat direction sheep breed had been formed. Compatibility of half mild wool to fat tail had been the first successful achievement in the history of sheep-breeding. This breed had been confirmed in 1949.Georgian half mild wool sheep is characterized by molting in spring. The average live weight of rams is 70 - 75 kg, of ewes – 45 - 50 kg. The average cutting of rams is 4.5 – 5 kg, of ewes – 3 – 3.5 kg. Output of pure wool achieves 60 – 65%.

The length of wool is 12-15 cm. Its mildness composes 50 - 56 quality. This breed is characterized by good meat peculiarities, but it is not fully perfect yet.

Georgian mild wool fat-tail sheep. This is such mild wool breed of sheep, which has fatty tail. It is formed by crossing of Tushuri sheep to Soviet merinos and mild wool Caucasian sheep.

This breed is confirmed in 1958. The average live weight of this original rams is 60 – 70 kg, of ewes – 42 – 50 kg. The average cutting of rams is 3.5 – 4 kg, of ewes – 2.5 – 3 kg. Wool is white and is characterized by sinuous merinos, its mildness reaches 60 - 64 quality, length – 7-9 cm. Output of pure wool composes 47 – 53%. Like Georgian half mild wool sheep, this breed is accustomed to difficult nomadic conditions of Georgia, too. But it is also characterized by molting, that's why it is not so popular in sheep breeding. The working on the completion of this breed is continuing.

Megruli goat

Two types of Megruli goat with dairy direction are bred in West Georgia: lowland goat and mountain goat. Live weight of mountain nanny-goat is 40 - 45 kg (max.50 - 60 kg), of billy-goat – 50 - 55 kg (max. 70 - 90 kg). Lowland goat is small but it's characterized by much more dairy.

Annually it gives nearly 300 - 400 kg milk by keeping in the pasture conditions during 6 - 8 month lactation, but the best of them give 800 kg milk with the 4% fat. Productivity is nearly 120%. It's characterized by the firm constitution; the hair is short and rough without baize, white or light grey in colour.

Georgian horse

The development of horse breeding in Georgia was caused first of all by living in the mountain conditions, then to frequent wars with foreigners and inside the country, but during peaceful period by agricultural activities. There were two types of horse breeding since ancient times in Georgia: stationary and pasture breeding.

From the diversity of breeds which were bred in those times, two types of species were preserved till present. These are Tusheti horse and Megruli horse. Besides them Javakheti horse breed group have been formed in Georgia.

Actions of Georgian National Association for Animal Production (GNAAP) for further development of livestock sector

GNAAP considers the following as the main priorities for livestock quick rehabilitation and development:
- to further establish and organize the breeding, dairy and other farmer associations structures which actively participate in monitoring of breeding and selection activities in the livestock sectors in the country;
- to start working to develop a central identification and registration system and data base for farm animals;
- to establish an animal gene bank;
- to organize extension activities.

The set-up of an organizational framework for the animal breeding structure is described in more detail:

A herdbook, (milk) recording service and AI-service should be established to have a firm base for cattle husbandry in Georgia. It's the base for increasing efficiency in cattle production, but it is also required to maintain biodiversity. The local breeds can be genetically improved by selection within breed, but also by using certain properties of the high productive Western breeds. To build a framework for animal breeding is a long time investment that needs support from the Georgian Government to be durable and successful. Cooperation with the other Caucasian countries (Armenia and Azerbaijan) is a way to make the investments easier to carry and the animal population large enough for animal genetic testing purposes. Also other donors than the State Government will be approached to add to the realization of this important and basic plan for organizing animal (dairy) husbandry in Georgia.

GNAAP is currently collaborating on the issues of livestock rehabilitation and protection and hopes to cooperate with other developed countries of Europe in this area as well. GNAAP is actively participating in setting up the Caucasus Regional Biodiversity Book

GNAAP is also working with the government on the set-up of an extension service.

The geopolitical state and political situation in Georgia conditions the necessity of creating a regional center for livestock development and extension activities for the South Caucasian countries located in Georgia. The situation in the livestock sector in the South Caucasian countries is directly similar. It is impossible to overcome the crisis in the livestock sector in these countries without appropriate scientific and practical support. The joint operations of South Caucasian countries will help to solve this problem.

Some thoughts about set-up of knowledge transfer points and training are as follows. Farmers and industry especially need support in the managements of their operations. Improving the milking techniques and hygiene, strengthening the feed base and feed production, improving feed quality and the quality of products for consumers are definitely main priorities in this area of knowledge transfer and training. How can this be achieved? The plans concentrate on this issue by proposing knowledge transfer points and training in these fields. The most efficient and perhaps cheapest approach is by creating for the time being for instance two centers of active knowledge transfer and training in two regions in Georgia. A possible spread to other regions is envisaged. Therefore these centers can be considered as pilots for a more wide spread approach in Georgia and the Southern Caucasus.

Other papers

Farm management under quota in small and large herd CEE countries

Marija Klopčič[1] and Jan Huba[2]

[1]*University of Ljubljana, Biotechnical Faculty, Zootechnical Department, Groblje 3, 1230 Domžale, Slovenia*
[2]*Research Institute for Animal Production, Hlohovská 2, 949 92 Nitra, Slovakia*

Introduction

Experiences in Western Europe learn that the introduction of the quota system will influence the management system of the farm. In this contribution a possible impact of the quota system will be described under a small scale farm structure and a large scale farm structure. Two new EU countries are taken as example: Slovenia and Slovakia and neighbouring countries.

Slovenia as example of smaller farms

With 20,250 km2 of surface area and nearly 2 million inhabitants, Slovenia can be considered as one of the smallest European countries. Slovenia is a distinctively Central European country, located between the Alps, the Adriatic and the Pannonian Plain. In spite of its geographically small size, it is a convergence point of a range of different landscapes: Alpine and Mediteranean, Pannonian and Dinaric, each of which has its own characteristics and unique features. Diverse natural conditions are reflected in the structure of the land utilisation, too. Slovenia is one of the most forested countries in Europe, since more than 60% of the country is covered with forest.

Figure 1. Map of Slovenia.

After the Second World War Slovene agriculture was developing under specific political and economical circumstances, which were fundamentally different to those in Western and Eastern Europe. Although agrarian reforms, designed to introduce collectivisation failed, the politicians kept stimulating the development of state agriculture until the end of the eighties, while at the same time neglected and hindered the development of private farming (until the 1970's).

Such circumstances, which lasted almost 50 years, considerably affected Slovene agricultural structure, production potential and market. Despite that, private farms preserved the predominant share in agriculture. The reasons lie in a great attachment of the Slovenian farmer to the land, his ingenuity and stubbornness, as well as in typical natural conditions. These conditions are in the major part of Slovenia less favourable for agricultural production and difficult to manage by big (state) farms.

Figure 2. Grassland in hilly areas of Slovenia.

The concept of the polycentric development of Slovenia also contributed to maintaining family farms. On the other hand, population density in rural areas made such a concept possible. Lately more and more attention has been paid to the role of farms and agriculture in the preservation of the presence of population in rural areas and of the cultural landscape. The role of agriculture is definitely more significant than it could be perceived merely on the basis of its share in the gross domestic product.

Structural changes before independence

During the period of land concentration in Western Europe, two agrarian reforms and deprivation of land (based on the Law on Land Maximum) took place in Slovenia. According to this law, the maximum size of privately owned land per farm was 10 ha of agricultural land

(20 ha for highland farms). The result was that land concentration on family farms for the following 45 years was hindered. Small-sized farms resulted in a low professionalism and productivity, that currently represents the biggest developmental problem in Slovene agriculture. In the period between 1981 and 1991 the average size of a Slovene farm hardly changed. Bigger changes happened after independence of Slovenia in year 1991.

Table 1. The number of family farms according to socio-economic type of farms.

Year	1991		1997		2000	
Type of farms	No.	%	No.	%	No.	%
Full-time	23,749	21.3	13,843	15.3	14,902	17.3
Part-time	54,891	49.2	27,452	30.3	30,333	35.1
Supplementary	22,122	19.8	39,473	43.6	32,570	37.7
Aged / other	10,784	9.7	9,691	10.7	8,531	9.9
Total	111,546	100.0	90,459	100.0	86,336	100.0

A distinct characteristic for Slovenia is a rather high percentage of part-time and supplementary farms, and a small percentage of full-time farms.

Full-time farms are farms on which all active household members (aged 15 to 64) work on the farm and are not employed outside the farm, i.e. the farmer, his wife, the heir and their children. The annual work unit (AWU) value must be at least 1.2.

Part-time farms are family farms on which members are active on the farm or outside it, or farms on which only household members employed elsewhere, retired persons and dependants work. The AWU value must be at least 1.2.

Supplementary farms are farms on which none of the household members (aged 15 to 64) work only on the farm. Only household members employed elsewhere, retired persons and dependants work on the farm. The AWU value must be at least 1.2.

Aged farms are farms on which only household members over 64 years of age live.

However, this structure is changing rapidly. In the period from 1991 to 2000 the number of total farms decreased with about 23%. The number of full-time farms dropped with about 9,000 and part-time farms with practically 25,000, while the number of supplementary farms increased with 10,000 farms. The number of aged farms has diminished.

Farming conditions

Because of predominantly hilly and mountain regions Slovenia is undoubtedly one of the Europe's countries with the least favourable conditions for agriculture. As much as three quarters of agricultural land lies in less favourable areas, and two third of the rural population live and work on farms with less favourable conditions. The natural conditions dictate significantly the exploitation of grassland and therefore the keeping of livestock. Cattle breeding prevails in the structure of agricultural production. Two third of all the cattle is bred and more than half of milk and the meat is produced in these less favourable areas.

In Slovenia arable land covers less than 30% of all agricultural land, and the share of the production on thise land represents approximately one quarter of the total agricultural output. Most of agricultural land in Slovenia is grassland (Figure 3), and because cultivated fields require rotation of fodder plants, it is understandable that the most important agricultural activity is ruminant production.

The farming conditions can be characterized as follows:
- Three quarters of agricultural lands with moderate farming conditions
- Many farms cannot increase agricultural production due to natural conditions and obstacles
- Land owning in Slovenia cannot be compared to land owning in the lowlands of Europe
- Natural allots and environmental sensitivity allow only one form of sustainable farming in some part of Slovenia.

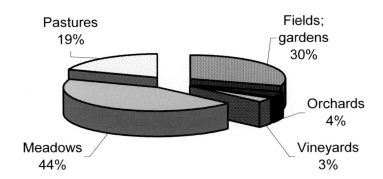

Figure 3. Structure of agricultural land in Slovenia.

Changes after independence in year 1991

After the independence in 1991 Slovenia decided to access the developed world as soon as possible. For that reason the choice was an open market economy that should consider our limited natural possibilities, i.e. the choice for a sustainable social market economy. The above objective has been entered into the Development Strategy of Slovene Agriculture that was passed by the Slovene parliament in 1993.

The openness of Slovene economy to the international market was confirmed in 1994 when we joined the World Trade Organization (WTO). It meant that our market opened to foreign economic products, import protections were eliminated, hence the price of domestic agricultural products fell. Farmers, agricultural enterprises and food processing industry encountered totally new conditions. They started to think about competitiveness of their production and products and introduced changes into their systems of livestock production and products. Conditions have been aggravated and the process of restructuring of production has started very quickly. The decision to access the EU brought negotiations on allowed production of milk, suckling cows and small ruminants on condition that environmental, ethological and other conditions that have already been accepted by the EU are considered. Slovene agricultural policy, advisory services and farmers are looking forward to adjusting to the new conditions. Farmers are aware that they will be successful in the EU as long as they can compete for prices and quality with the farmers in the EU.

After the accession to the European Union (EU) the Slovene livestock production encounters altered conditions. The production should adapt to the values and quality that have been achieved during the negotiations with the EU. The present milk purchase is nearly equal to the negotiated quotas. In the coming years the number of dairy cows and number of farms with dairy cows will decrease. Livestock production remains the most important sector in the

Slovene agriculture, although its share in the volume of total agricultural outputs has recently dropped to below 50%. The biggest share in livestock production is represented by cattle husbandry, followed by pig and poultry breeding. Other livestock production, like small ruminants breeding and horse breeding is less significant and smaller in volume.

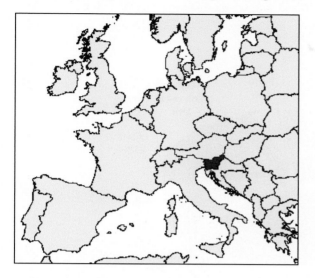

Figure 4. Slovenia as part of Europe.

Cattle production

Cattle husbandry is the predominant production orientation owing to our natural conditions (i.e. large proportion of grasslands). It has been evolving primarily on family farms. The volume of cattle breeding on agricultural holdings is relatively modest, and in recent years it has decreased even more as well as the number of cattle (Table 2).

Table 2. Number of cattle, suckler and dairy cows in Slovenia.

Year	Cattle - total	Dairy cows	Suckler cows	Cows –total
1980	558,144	146,807	79,310	226,117
1985	577,206	166,252	56,299	222,551
1990	546,048	161,992	58,274	220,266
1995	495,535	137,532	74,786	212,318
2000	493,670	135,594	69,189	204,783
2002	481,514	113.599	70.837	184.436
2004	450,226	111,251	86,661	197,911

Cattle production presents 40% of the final agricultural production in Slovenia (Figure 4). Recently the rate of small ruminant breeding and production is steadily rising It is therefore clear that self-sufficiency of beef and veal comes to almost 100% and in the case of milk, there is 20% surplus in Slovenia (Table 3).

The milk surplus is mainly the consequence of lower milk consumption per capita (210 litres) in comparison with EU countries. On the other hand, beef consumption in Slovenia is 22 kg per capita, which is higher than in the EU countries (Table 3). We hope that better supply and competitive prices as well as advertising activities will contribute to higher milk consumption after the access to the EU and that we will find buyers on the European markets.

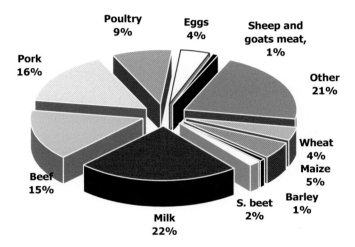

Figure 4. Value of agricultural production in %.

Table 3. Consumption per capita (kg/yr) and levels of self-sufficiency (%).

	Consumption per capita (kg/yr)		Levels of self-sufficiency (%)	
Year	1995	2000	1995	2000
Meat and entrails	95.8	94.3	96.0	92.5
Beef	27.6	21.8	94.1	97.2
Pork	36.0	38.7	83.0	76.2
Poultry meat	26.1	29.1	119.2	109.6
Sheep and goat meat	0.3	0.6	92.6	99.9
Milk – total	210.3	225.6	114.9	120.2
Milk – fresh milk and milk beverages	118.2	129.4	127.5	125.0
Cheese and curd	8.7	10.0	97.3	107.9
Butter	0.9	1.1	130.7	159.9
Eggs	9.9	11.5	106.7	96.2

Changes in cattle husbandry due to quota system

Because of the importance of cattle husbandry, the negotiations for allowed amounts of beef production and especially milk production were very important for Slovenia. Field surfaces should be used for rotated grass and clover production to maintain the fertility. Therefore a lot of fresh fodder is produced that could be consumed only by ruminants, especially by cattle. It will be interesting to compare the past and present production and to estimate our opportunities after the accession to the EU. Special attention should be paid to production prices that together with allowed subsidies should enable the competitiveness of our products on the EU markets. This objective could be achieved only with significant structural changes.

The production goals before access to EU were:
- to increase milk yield of dairy cows;
- to improve milk quality;
- to rise the number of suckler cows;
- to improve the inventory of small ruminants.

Negotiations for production of cattle and small ruminants with the EU led to allowed amount of production and rights to pay subsidies for slaughtered animals, which represent the base for the planning of animal production in Slovenia (Table 4). There are no limits in pig and poultry production.

Table 4. Allowed amounts of production.

Product	Unit	Amount, May 2004	Amount, December 2005
milk – total	tons	560,424	560,424
reserve for the year 2006	tons	16,214	16,214
sale to diaries	tons	467,063	515,743
direct sale	tons	93,391	44,681
suckler cows	number	86,384	86,384
small ruminants	number	84,909	84,909
subsidies – bulls, oxen	number	92,276	92,276
slaughter subsidies	number	196,989	196,989
adult animals	number	161,137	161,137
calves	number	35,852	35,852

Changes in management due to quota system

We expect, that the production goals will be influenced by the milk quota system. First the milk quota system will be described shortly:
- The quota and premium system for milk and suckler cows was introduced in year 2004/2005 (this is a trial period);
- A reference year was the milk production in trial year 2004/2005;
- In April 2005 the milk quota was officially in use;
- The experience in old EU countries shows, that the adaptation of agriculture and institutions to the new situation is a very complex process and takes much time and efforts.
- The situation in Slovenia is especially complex, because of the presence of relatively small farms.

- The adaptation is necessary to avoid large penalties by exceeding the assigned quota amounts and to utilise the premiums in an efficient way.
- The most important is to instruct consultants and institutions how to deal with EU instruments to work on a perspective for the future.

At the moment milk delivery quota consist of 515,743 tons and the quota for direct sale 44,681 tons with 4.13% of fat content. The amount of purchased milk in year 2004 was 486,010 tons and in year 2005 was about 500,000 tons. In next two years we don't yet expect exceeding the assigned quota amount. This may change in years to come. Then overproduction may result in a:

- reduction in production on farm level;
- possible consequences:
  o breeders' complains / political problem;
  o diminished number of dairy cows;
  o less animals/ha;
  o lower use of feeding mixtures;
  o less fertilizer use.

Data in Table 5 show that significant structural changes have recently occurred in dairy husbandry. The number of farms (herds) that produces market milk has been diminished. The number of cows diminished too but not significantly because herd size increased. Due to increased share of Black-and-White cows and to higher production ability of Brown and Simmental breed cows, the total amount of purchased milk increased as well as purchased milk per farm and per cow (Table 5).

Table 5. Number of herds and cows and amounts of purchased milk.

Year	Herds, No.	Cows. No.	Total milk production, kg	Milk production per cow, kg	Milk production per herd	Fat, %	No. of cows /farm
1980	55,533	150,694	303,831,000	2,016	5,471		2.71
1985	58,194	175,696	352,454,200	2,120	6,063		2.86
1990	43,656	161,992	359,184,200	2,217	8,228	3.74	3.53
1995	30,040	132,532	388,394,400	2,968	12,942	3.92	4.36
2000	16,869	117,775	447,831,000	3,758	26,516	4.10	6.79
2002	12,589	113,599	473,500,000	4,168	37,612	4.13	9.02
2003	11,289	112,484	484,200,000	4,305	42,891	4.14	9.96
2004	10.133	105,000	486,009,740	4,796	47,963	4.16	10.36
2005	9.539	106,000	502,000,000	4,736	52,626	4.15	11.11

Due to improved production technologies, better feeding regime and production abilities of cows the milk production per cow will increase. Therefore number of dairy cows is expected to diminish to 90,000 in 2010. It is anticipated that a part of quota for indirect purchase can be moved to the quota for purchase by dairies. Hence it is estimated that farms will produce over 520,000 tons of purchased milk, which will be over 5,500 kg per cow and more as 65,000 kg per farm. Consequently the average herd size will increase to 12.5 cows or even more, while the number of farms with low number of cows will diminish. Milk production will be present only on relatively big farms.

In coming years the question should be solved how to achieve the allowed annual slaughter of animals as well as high quality with allowed number of sucklers and allowed quotas of produced milk that limit the number of cows. According to the data on number of inseminations (Figure 5) it could be stated that Slovene farmers rear about 180,000 cows that are in general involved in the production process for 5 years.

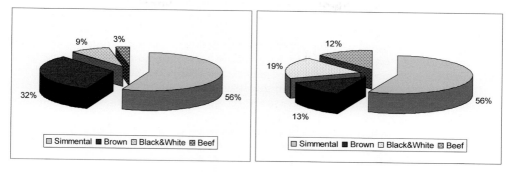

Figure 5. Artificial insemination of cattle by breed in year 1985 and year 2005.

The inclusion of Holstein-Friesian cows into milk production and crossing of Brown breed with dairy American Brown Swiss causes that the amount of good calves for meat production has been diminished. Therefore, the quality of pure breed fattened slaughtered cattle deteriorated, but consumers ask for beef and veal of the best quality. Quality is paid well, good quality calves for fattening reach good prices and breeders ask for them. Hence breeders of Holstein-Friesian and Brown cows have become interested in industrial crossing. Last year above 22,000 Slovene cows were inseminated with bulls of meat breeds. Due to the fact that calves of Simmental cows are more suitable for fattening than Brown and Holstein-Friesian calves, it is expected that for industrial crossing Brown and Holstein Friesian cows will be used. From data we can see, that breeders dedicate one third of the lowest producing Brown and Holstein-Friesian cows for industrial crossing. This crossing is popular to be used as a suckler cow. Such crossings maintain the quality of young cattle which Slovenia used to be famous for.

Figure 6. Simmental and Brown cows on the mountain pasture.

In cattle production the number of suckler cows is rapidly increasing. These are mostly Simmental, Brown and combined type of cows. On those farms where suckler cows are kept, economic crossbreeding with beef breeds is quite intense: Limousin, White/Blue Belgian breed and Charolais. Crossing of Simmental and Limousin breeds is mostly performed on farms oriented to ecological cattle production of beef and veal. Joining EU, the assertion of subsidies as well as higher demands for ecological beef production will definitely increase the number of young steers, too. In future, economic crossing will, no doubt, contribute to the improved slaughter mass of young fattened animals and better slaughter quality.

The same situation is expected also in the case of ecologically produced milk, and milk products. Furthermore, farmers will have to respect ethological principles when building new stables.

We expect the following management practices to be employed at start of quota system:
- improve efficiency of farm operation;
- more emphasis on quality of products to increase the price;
- improve quality of crop and grassland;
- improve animal health status (mastitis, SCC, fertility…);
- attention for feeding regime;
- more interest in machine societies – interacting help with machines.

Figure 7. What is the future for Slovenian farmers?

Besides management practices in the short run, we think that strategy development in the long run should stimulate farmers to consider their future options under the new circumstances and supply management regulation. The following strategies are realistic options:
- Suckler cows.
- Beef production will be increased.
- Small ruminants (sheep and/or goats).
- Pig production.
- Horses for sport and recreation.
- Recreational activities.
- Supplementary activities as agro tourism, cottage industry, services with machines.
- Special local products.
- Management of forests.

The quota system will have impact on farm management. This may not be the case so much in the first year after the introduction of quota. The awareness concerning the new situation is not yet matured. This takes time. But in years to come, the farmer has to adjust his production volume to the quota amount assigned and/or will look for new ways to develop the farm further. Quality of farming will increase in importance in comparison to the traditional strategy of growth of milk volume.

Figure 8. Knowledge transfer to the practice is very important.

Slovakia as example of large farms

As example of an agricultural structure with larger farms Slovakia and neighbouring countries are taken. The large size farm structure can be described as follows:
- More than 100 cows/farm.
  o Slovakia 170 cows/farm.
  o Hungary 310 cows/farm.
- Mainly free stalls with milking parlours.
  o Slovakia 60% of farms.
  o Hungary 70% of farms.
- Modern technologies (last 10 years).
- Holstein breed prevails.
  o Czech Republic 53% of cows.
  o Hungary 83% of cows.

Management decisions at start of Quota System

We may want to decrease the costs on milk production. This can be achieved by:
- improvement of fertility traits:
  o calving interval in Hungary –   434 days.
                   in Slovakia –   425 days.
- increase of longevity (2-3 lactations now);
- decrease of costs in heifer rearing (pasture);
- improvement of roughage quality;

We may increase the revenues from the dairy cows sector. This can be achieved by:
- improvement of milk quality;
  o decrease of somatic cell count;
  o increase of milk protein content.
- increase of beef production per cow;
  o Hungary 260 kg live weight;
  o Slovakia 294 kg live weight;

Changing of animal breeding goals

As a result of the quota system we may expect more emphasis on:
- protein content;
- healthy status of dairy cows;
- longevity traits;
- fertility traits;
- fitness;
- beef traits.
  As example we present the economic value of the Austrian Simmental:
  Milk (37%)Beef (18%)Fittnes (45%)

Increasing performance

Challenges and problems facing the new situation can be characterized as:
    Breeding   Increase of performance

Solutions

- decrease of dairy cows numbers (higher production level);
- increase of interest in dual-purpose and local breeds;
- purchase of quota (market price of quota?, number of farms diminishing; unemployment in rural areas);
- different solutions to deal with milk surplus above assigned quota.
  An option to improve the performance of the farm business under quota is to work with quality programs. We think about:
- further improvement of the overall quality standards of milk and meat;
- proper maintenance of the milk and stable equipment;
- proper use of medicines;
- proper additives in feeds;
- proper mineral and pesticides management.
  In summary: Best management practices will be advocated!

Conclusions

- The number of dairy cows will diminish.
- Health condition of dairy cows should be improved in order to diminish veterinary costs.
- Structure of costs will change.
- Greater emphasis will be on the quality of products, better fertility, longevity of animals and health of animals.
- With very large farms, in first years adaptation to assigned quota is possible without big problems.
- Then new goals have to be set.
- Improve quality of farm management.

Development of dairy cattle farming systems in CEEC: Experiences from consultancy visits

Arunas Svitojus

International Charitable Foundation for Baltic Countries, S. Konarskio 49, 03123 Vilnius, Lithuania

The changes of management and use of dairy cattle in livestock farming systems

From 1990 on the transition from the centrally planned economy to the market economy in Central and East Europe countries caused a crisis that, on its part, resulted in a reduction of the number of farm animals, decrease in productivity and a down fall in the animal production industry. The difficult situation has been developed in many branches of livestock farming. The collective farms have been ruined and dismantled, the big number of livestock have been slaughtered, in spite of the fact that reduction of cattle was significant and the great part of countries have been not self-sufficient in livestock production.

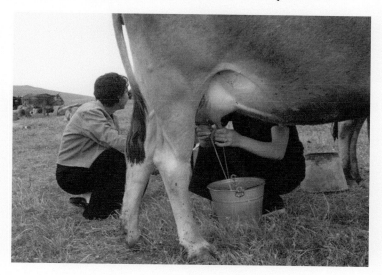

Figure 1. Hand milking most popular in Eastern Europe.

Since the dissolution of the Soviet Union in 1991 all of the 15 former Soviet republics to a regions with the specific geographical and economical developments, such as Caucasus countries Armenia, Georgia, Azerbaijan, Baltic countries Estonia, Latvia, Lithuania, Central Asia countries Kazakhstan, Kyrgyz, Tadzhikistan, Turkmenistan Uzbekistan and East Europe countries Belarus, Moldavia, Russia, Ukraine, have dismantled their Soviet-style economies.

Traditionally the national milk sector in Baltic States was export – oriented. Changes in the trade pattern with Russia and other CIS states and establishment of new partner relations in other markets resulted in the reduction of dairy exports. In 2002 it comprised 32 % of total export.

After accession of three Baltic Countries to the EU in 2004, there are expected liberalization of trade in agriculture goods and the stabilization of livestock production and marketing.

Figure 2. Cows on the pasture.

Figure 3. Holstein-Friesian cows.

Now the situation started showing improvement indications; to a certain degree, due to the increase in the production volumes of domestic animal husbandry products. The quality of domestic products has also improved.

In the Strategy for Agricultural and Rural Development in Lithuania, the dairy sector has been acknowledged as a priority branch of agriculture. In 2002 milk production made up 19.7 percent of the total value of agricultural production, whereas the export of dairy products made up over 32 per cent of the total exports of agricultural and food products.

Changes in the milk production have been caused by the period of transition from a planned to market economy. In 2001, in comparison to 1998 the number of cow reduced by 12 %. In 2002 the number of cows kept stable. Individual farmers keep about 92 % of cows and produce 91 % of total milk production.

Currently, there are three main types of livestock farming systems on which domestic animals can be raised:

- Cooperatives replacing former collective farms;
- family farms;
- small holders.

In 1990 – 2002 the number of farm animals sharply declined. Cattle declined by 69%.

In 2002, the average milk yield per cow reached 3,946 kg. The average milk yield of controlled cows (30% of the total number of cows) amounted to 5,015 kg.

Small dairy farms take a dominant position in Lithuania. In 2002 milk was produced by 197, 4 thousand farmers and 176 agricultural partnership farms. On average in 1 farm is kept 2.1 cow. Only 1 % of all farms keeps 10 and more cows. They keep 16 % total cow populations.

Table 1. Changes of animal population (thousands) in Baltic Region in 1991 and 2004.

	Cattle		Of which: Cows	
	1991	2004	1991	2004
Lithuania	2321.5	812.1	842.0	448.1
Estonia	708.3	249.1	264.3	114.9
Latvia	1439.0	371.0	535.0	186.0
Total	4468.8	1432.2	1641.3	749.0

Changes in the milk production have been caused by the period of transition from a planned to market economy. In 2001, in comparison to 1998 the number of cows reduced by 12%. In 2002 the number of cows kept stable. Individual farmers keep about 92 % of cows and produce 91 % of total milk production.

In the Strategy for Agricultural and Rural Development in Lithuania, the dairy sector has been acknowledged as a priority branch of agriculture. In 2002 milk production made up 19.7 percent of the total value of agricultural production, whereas the export of dairy products made up over 32 per cent of the total exports of agricultural and food products.

Livestock production in Central Asia and Caucasus countries is distinguished by its richness and variety. Different animal species are distributed in various agro-ecological zones. Milking cows, for example, are mainly found in irrigated croplands near industrial centers; beef cattle on pastures in mountainous areas; sheep in deserts; rams and horses on the foothills and mountains. The distribution of animal production systems is dictated by feed availability and climate.

The situation created in former Soviet Union countries during the process of transition from the centrally planned to the market oriented economic system imposed the need for adjustments and reforms- in all system. The fall in consumers incomes and rise of prices has fallen down the real demand for animal husbandry products. That has reflected in dairy sector too.

The production level of animal food products does not meet the population requirements. For example, under the current conditions Armenia is still far from securing its own needs for foodstuff. Until recent ten years domestic production of milk and dairy products was enough to meet less than half of local demand. Armenia - Besides traditional products, also on sale are yogurt prepared of sheep and buffalo milk. A certain progress is in milk goat rising.

Figure 4. Cows in less favoured areas.

Figure 5. Local breeds and crossing.

The importance of products relates to particular regions, depending on geographic, social and economic status and management of natural resources.

Currently, there are three main types of livestock farming systems on which domestic animals can be raised:

- state-run agricultural cooperatives;
- cooperative replacing former collective farms;
- family farms.

The animal breed evolved from years of thorough selection in its natural habitat. Consequently, it is highly adapted to the harsh climatic conditions, but also to local management strategies.

In the mid-1990s, the national research organizations started to collaborate to better understand the socioeconomic processes and changes brought about by the new economic reforms. They developed strategies to overcome production problems dairy farmers had to face.

The planning of activities primarily focused on products that had an opportunity on the local markets. Major activities were framed into the following research components:

- Markets and socioeconomics of production systems;
- On-farm interventions for productivity improvement;
- Efficient livestock production through adequate flock management practices integrating nutrition, reproduction, breeding, animal health and quality of products.

Caucasian countries have a great prospective potential to use its rich natural complexes for different purposes, for instance, for agriculture. Specific natural and economic conditions greatly affect Caucasian agriculture. About 55% of people are employed in this sector.

Figure 6. Pasturing of cattle in Central Asia and Caucasus.

Figure 7. Feeding of dairy cows.

Figure 8. Moving cattle from one grazing area to the other.

Nowadays small and medium scale livestock farms compose the main part of the household agriculture. There exist large farms mainly with dairy cows which have arable lands for providing feed base for animals. On farm level, the management of farms is problematic in Caucasus. It's necessary to implement new effective management.

Figure 9. Guarding of the cattle.

In Caucasian countries the production level of animal food products does not meet the population requirements. For example, under the current conditions Armenia, Georgia and more or less Azerbaijan are still far from securing its own needs for foodstuff. Until recent ten years domestic production of milk and dairy products was enough to meet less than half of local demand. Besides traditional products, also on sale is yogurt prepared of cow and buffalo milk.

Table 2. Changes of animal population (thousands) in Caucasus Region in 1991 and 2004.

	Cattle		of which: Cows	
	1991	2004	1991	2004
Armenia	640.1	478.7	250.9	262.1
Azerbaijan	1826.2	2293.6	733.7	1107.5
Georgia	1298.3	1242.5	551.7	728.0
Total	3764.6	4014.8	1536.3	2097.6

During the last decade in East Europe essential changes have occurred in agrarian sector, conditioned, as it was already mentioned, by the transition from centrally planned to the principles of market economy relations. In this connection the enterprises and organizations, which were occupied by producing products were exposed in the essential reconstruction and lost big quantity of animals.

In Belarus negative tendencies in livestock rising are caused by a number of economic and organizational characters. Because of price disparity agricultural enterprises cannot buy enough forage of proper quality population of animals decreased about 50%.

In 1996-98 an intensive fall of cattle heads in livestock was noticed, but after 1998 this process somewhat slowed down and in some branches of livestock the increase of cattle heads was noticed (for example, in pig and poultry breeding). In farms cattle heads stayed relatively stable, but insignificant, comparing to agricultural enterprises. The same tendencies were noticed in private subsidiary farms of population.

Table 3. Changes of cattle population (thousands) in East Europe Region in 1991 and 2004.

	Cattle		Of which: Cows	
	1991	2004	1991	2004
Belarus	6975	3970	2360	1618
Moldova	1061	405	450	276
Russia	57043	22988	23857	10252
Ukraine	21083	8251	6192	4284
Total	94198	35614	32859	16430

The cattle-breeding in Moldova was and remains the main source of protein for valuable people's nutrition, source of raw materials for light and food industry, and also for organic fertilizers for plant cultivation. In conclusion, the cattle-breeding is one of the leading branches of Republic's agro-industrial complex.

In the period of 90[th]s, up to 80% of the whole volume of cattle-breeding production was produced into the system of scientific-production association "Kolkhozjivprom". It contained specialized kolkhoz' farms, industrial type enterprises for milk, meat, poultry products production, heifers nurture for herd reproduction, and also enterprises for storing-up and realization cattle breeding products.

In Russian Federation the main branches, which play an important role in producing livestock products are dairy and beef livestock, pig breeding, sheep breeding and poultry breeding. Privatization process of state property affected all branches of livestock practically in the same way. Together with that the total number of agricultural enterprises slightly reduced: dairy livestock by 17.5%. Practically the number of farm households in dairy (31-32 thousand) and beef livestock (1.3 - 1.5 thousand).

It seemed the tendency of increase in the number of private subsidiary farms in dairy livestock (by 18.3%). In the enterprises which produce milk, beef and pork up to present exist large complexes built during Soviet period.

During the 90-s of the past century there was an essential decrease in the number of cattle heads of all species practically in all categories of farms (except private sectors). Mainly this process was conditioned by the following reasons:

- Quick decrease of the influence of government (first of all - economical) on keeping of agricultural animals and on producing of livestock products.
- Instability of macroeconomic ratio of country's development during transition period, which contributes the appearing of price disparity on livestock products and expenses on its producing (first of all - prices of energy resources, agricultural techniques and equipment).
- Quick fall of customers' abilities in the majority of country's population.
- Expansion of imported products on market of Russia (frequently by damping prices).
- Undeveloped infrastructure of producers of agricultural products with irregularity of its development through the regions and zones of the country.

During the last ten years in Ukraine, the cow stock decreased 3.7 times in the agricultural firms, the average annual yield of milk being 1.5 – 2 times, the fodder consumption was reduced 12 - 22% at the same time, which made the milk production unprofitable. Improvement of feeding which is evidenced by the increase of fodder consumption in 2001 for the current number of cattle caused an increase of the average annual yield of milk. This helped the homesteads reach nearly the zero level of profitability. That is, if the milk price and cost parity were preserved at the level of 2001, and with the current number of cattle (1.68 million head), the fodder consumption increase would help raise the average yield of milk and profitability.

Absence of profit from milk production is the prime cause of the livestock decrease. The losses are primarily caused by high prices for energy sources.

Livestock production in Central Asia and Caucasus countries is distinguished by its richness and variety. Different animal species are distributed in various agro-ecological zones. Milking cows, for example, are mainly found in irrigated croplands near industrial centers; beef cattle on pastures in mountainous areas;

Table 4. Changes of animal population (thousands) in Central Asia Region in 1991 and 2004.

	Cattle		Of which: Cows	
	1991	2004	1991	2004
Tajikistan	1377.0	1278.0	580.8	642.3
Uzbekistan	4580.0	5878.0	1856.0	2556.0
Kyrgyz	1190.0	1034.9	518.6	548.2
Kazakhstan	9755.7	4854.6	3367.1	2272.3
Turkmenia	829.0	1969.9	447.8	1007.6
Total	17731.7	15015.4	6770.3	7026.4

In recent years in Kazakhstan the cattle and poultry productivity increased greatly. Thus, in 2000 the average annual milk yield per cow in all farm categories exceeded the 1998 level by 194 kg. Due to the recent growth in the cattle and poultry number and productivity, the gross milk yield increased. Animal breeding branch of agriculture had hard times during the period of reforms in Kazakhstan. One of the main reasons of this situation is the considerable reduction in the number of domestic animals and poultry in 1991-1998. In 1999-2000 the situation was improved: the number of cattle in 2000 against 1998 increased by 3.7%, sheep and goats – by 4.7%, pigs – by 20.6% and poultry – by 15.8%.

In Kyrgyz in most cases, all these economic entities are engaged both in agriculture and livestock industries. As a rule, in all private and peasant farms 3-4 species of cattle are ranched or bred, there are large horned livestock, sheep or goats, agricultural poultry, horses (especially in mountain regions) more often. In large cooperative and state farms basically 1-2 species of cattle are bred or ranched.,

In 1990 in the republic it was totaled 1.2 million heads of cattle. By the beginning of 2002 number of cattle was reduced up to by 986.1 thousand. In livestock number though for the last 3-4 years a cattle breeding was somehow stabilized and the tendency on increase was outlined. Reduction in livestock in many respects has been connected to change and reformation of socioeconomic and industrial systems, during the early period of a transitive stage – to farmers inability to manage in the changed conditions, during the late period it has connected to economic calculation of profitableness of this or that branch of animal industries.

In 1990-s in Tajikistan cattle breeding took one of the prominent places among other industries by quantity and productivity. Sheep breeding and goat breeding in mountainous conditions are very important branches of stock breeding in the republic. They were always traditional. The production of these branches are in popular demand of population and that's why their meat and fat cherish are valuable, than beef and other types of meat.

At the present time many lease and farming facilities use manual labor (the collection of harvest, primary processing, storage), milking, preparing of provender, haircuts and so on. For the last decennial events as a result of epy civil war, economic and political crisis, supply of the population with products of agriculture, particularly with products of stock breeding sharply fell down.

At the beginning of 2004 in Turkmenistan there was number of cattle - 1.97 million heads. Nowadays the more than 94% of cattle are in private sector.

The cattle located in agricultural farms (farmers associations) handed over sublet maintenance, i.e. in livestock breeding sector there are the production systems such as tenet relations. Farmers associations (former kolkhozes) conclude the contract with shepherd's teams. The tenant accordingly to the contract conditions gets the certain part of breed issue and products. All expenses on cattle maintenance bears the tenant.

The farmers association apportions pastures (free of charge), watering vessels and farms, if it is necessary, helps with forage and cattle transportation, marketing the products.

In cattle breeding on the tenancy condition tenants get watered lands for forage producing on the basis of 0.5 hectare on each cattle head. Tenants accordingly to the task must receive from 100 cows, 80 calves the milk on 2000 kg from each cow. Under the agreement conditions 60% of issue and all milk are remained for tenants.

In Uzbekistan the stock-breeding of the republic is presented basically by cattle breeding, sheep breeding and fowling.

The cattle breeding is highly developed in irrigated zones. The cattle-breeding farms are concentrated basically in collective and cooperative farms concerning with production of cereal crops and cotton.

At present the main part of livestock is kept by householders and private farmers.

The production system in the country is all basically self-provided, only greater dependency presents for veterinary service. Big part medication, veterinary preparation, vaccines and diagnose are imported.

After 15 years of reforms former Soviet republics have real difficulty meeting their husbandry (agricultural) needs concerned such as restructuring of farming systems and lack of information in Soviet Union countries induced these countries to connect to international organization.

Table 5. Changes of cattle population (thousands) in Post Soviet Union Region in 1991 and 2004.

Region	Cattle		Of which: Cows	
	1991	2004	1991	2004
Baltic	4468.8	1432.2	1641.3	749.0
Caucasus	3764.6	4014.8	1536.3	2097.6
East Europe	94198.0	35614.4	32859.0	16430.4
Central Asia	17731.7	15015.4	6770.3	7026.4
Total	120163.1	56076.8	42806.9	26303.4

The future courses of action with the aim to restructure the cattle sector are:
- try to achieve vital commercial farms;
- modernization of production capacities and agro-technological and agro-economic procedures;
- increase in the role of farmers on the agricultural product market;
- impact on the changes in the agrarian structure and production technology in order to increase competitiveness of agricultural production;
- increase in efficiency of administrative and special services and associations in agriculture;
- stimulation of personnel training;
- stimulation and development of ecological and traditional agriculture that allows the survival of a relatively rich animal world.

All this will considerably affect production systems in cattle production. Countries are still experiencing problems in the transition of the animal production sector towards the market economy systems.

There is a need to keep higher milk productivity!

Figure 10. Family help in dairy husbandry.

Economic weights for milk yield traits under the quota restriction in Slovakia

Jan Huba, Jan Kica, Jozef Daňo, Emil Krupa, Marta Oravcová, Ladislav Hetényi

Research Institute for Animal Production, Hlohovská 2, 949 92 Nitra, Slovakia

Summary

Economic weights of milk yield traits were calculated as partial derivates of the profit function using closed herd model for population of Holstein, Slovakian Simmental and Slovakian Pinzgau cattle. The calculations were carried out with and without quota for milk and milk fat. All revenues and costs connected with one generation of progeny of selected parents were discounted at a discounting rate of 2% to the birth date of progeny. The following economic weights with quota per standard female unit were calculated for Holstein, Slovakian Simmental and Slovakian Pinzgau breeds: milk yield with constant fat and protein contents 7.17, 7.91, 8.99 SKK/1 kg, fat content 1828, 1694, 1899 SKK/1%, protein content 11 395, 8996, 8489 SKK/1%, milk carrier 0.298, 0.51, 0.751 SKK/1kg, fat yield 32.99, 31.35, 42.20 SKK/1 kg, protein yield 158.27, 166.59, 188.65 SKK/1 kg (1 EUR=41 SKK).

Introduction

Milk payment systems vary across the countries. As a result production indices for selection of dairy cattle are different, too. In the indices the breeding values of each traits are weighted by economic weights. Generally, economic weights are calculated as partial derivatives of the profit function. All the important parameters of the bio-economic system under analysis must be involved in the profit function (Wolfová *et al.*, 1996). Economic weights were firstly calculated for milk (without fat) and fat yield (Adelhelm *et al.*, 1972, Henze *et al.*, 1980). Afterwards, economic weights for protein yield were calculated (Groen, 1989; Wolfova *et al.*, 1992; Solkner *et al.*, 2000). Restriction under quota (as could be found in Groen, 1989) presents a new approach in calculations of economic weights and constructions of selection indices. For Czechia with production and economic circumstances similar to Slovakia, Wolfová *et al.* (2001) carried out calculations with and without milk quota. In Slovakia, restriction on milk production as a result of entering the European Union comes into force on 1 April 2004 therefore new breeding strategies under changed situation need to be adopted.

The aim of the work was to calculate the economic weights with and without quota for milk performance traits for Holstein, Slovak Simmental and Pinzgau breeds and to compare both calculations.

Material and methods

For the derivation of the economic weights, a bio economic model of a closed herd was used which included the whole integrated production system of a dairy or dual-purpose breed. The total discounted profit for this herd was calculated as the difference between all revenues and costs that occurred during the whole life of animals born in the herd in one year and that was discounted to the birth year of these animals:

$$Z_T^0 = Z^0 S_{StFU}$$

$$Z^0 = \sum_k N_k (R_k q_{R_k} - C_k q_{C_k})$$

where:

Z_T^0	= total discounted profit in the population of the given breed (closed herd)
S_{StFU}	= number of standard female units (StFU = one cow place occupied during the whole year)
Z^0	= discounted profit per StFU
N_k	= average number of animals in category k per StFU
$R_k C_k$	= average revenues and costs, respectively, per animal of category k
$q_{Rk} q_{Ck}$	= discounting coefficient for revenues and costs, respectively, in category k

The discounting coefficients for the revenues were calculated by the following equation:

$$q_{Rk} = (1 + u)^{-\Delta t_{Rk}}$$

where:

Δt_{Rk}	= average time interval between the birth of animals of category k and the time of collecting revenues
u	= discounting rate (expressed as a fraction)

The not discounted profit (that means the average profit per year in the whole balanced system) was calculated by setting $u = 0$ so that all q´s took the value 1.

The discounted economic weight of a given trait i was defined as the partial derivative of the total profit function for the closed herd with respect to the given trait whereby all traits were assumed to take their mean values:

$$a_i = \{ \partial Z_T^0 / \partial x_i \big|_{x=\mu} \} / S_{StFU}$$

where:

x_i = value of the trait i under consideration

x = vector of the values of all traits (dimension of x = number of traits)

μ = vector of the means of all traits

Detailed definitions of all evaluated traits and a complete description of the method and the individual models used for the calculation of economic weights can by found in Wolfová and Wolf (1996). The computer programme EW (Wolfová and Wolf, 1996) was used for all calculations after some minor actualisation of the programme.

For milk production traits, the programme offers two ways for their definition. They can by expressed either as amount of milk carrier, fat yield and protein yield (all three traits in kg) or as milk yield (kg) with given fat and protein contents (both in %). The situation with milk quota with reference to fat content and the situation without milk quota were calculated. A decrease in the number of the standard female units was assumed as a response to exceeding the limit for the quota.

As a response to an increased productive lifetime of cows, the following three scenarios were taken into account to calculate the economic weight of cow longevity.

Input parameters

a) Production and reproduction traits involve dairy traits, fertility traits, their relations and herd replacement parameters. The calculations of actual production system parameters were based on milk recording data of year 2003. Cow culling parameters were calculated from data of cows culled in previous year. The data were obtained from the State Breeding Institute Bratislava. Input values were adjusted for standardized lactation of 305 days, average calving interval and average age at first calving. Coefficients for 2nd and 3rd lactations were calculated from 1st lactation parameters. With considered breeds, production and reproduction parameters were predicted for year 2008 taking into account all expected changes.

Table 1. Expected biological parameters for production system in 2008.

Parameter (unit)	Slovakian Simmental	Slovakian Pinzgau	Holstein
305- day milk production in 1st lactation (kg)	5,400	4,500	7,200
Milk protein content (%)	3.35	3.30	3.25
Milk fat content (%)	4.10	4.00	4.00
Average number of lactations per StFU	3	4	3.
Maximum number of lactations per cow	10	10	10
Calving interval (days)	405	395	415
Genetic standard for milk production f(kg)	500	350	650
Age of heifers as 1st breeding (days)	550	610	480
Coefficient of repeatability for milk production	0.5	0.5	0.5
Discounting rate (% per year)	2	2	2
Proportion of calves born alive from all calves born	0.95	0.95	0.95
Proportion of calves alive at ERPC from calves born alive	0.92	0.92	0.92
Proportion of heifers calving from calves alive ERPC	0.42	0.42	0.42
Proportion of bulls fattened from calves alive at ERPC	0.5	0.5	0.5
Proportion of heifers culled not for low milk yield from heifers calving	0.1	0.1	0.1
Milk production in lactation 2nd : 1st	1.16	1.2	1.1
3rd : 1st	1.23	1.28	1.13
4th : 1st	1.25	1.32	1.13
5th : 1st	1.25	1.34	1.1
6th : 1st	1.24	1.33	1.08
7th : 1st	1.23	1.31	1.06
8th : 1st	1.22	1.27	1.00
9th : 1st	1.20	1.26	0.92
10th : 1st	1.16	1.25	0.92

b) Economic parameters – costs
Changes in prices for concentrate and roughage feed were calculated with respect to their changes within the European Union. With considered breeds, the expected average year feed ration was calculated from production parameters predicted. Moreover, the expected variable

costs for milk production were predicted with respect to technology changes and expected increase in labour costs.

c) Economic parameters – milk prices

The expected milk price for year 2008 was calculated with respect to expected changes in milk price at world market. Also, actual and future farm prices in the European Union member countries were taken into account, mainly milk prices in period 2004-2008. In accordance to the European Union milk price premium that is assumed for period after 2007 the additional payment 1 SKK/kg was added to milk carier price.

Table 2. Expected economic parameter for production system in 2008.

Parameter (unit)	Slovakian Simmental	Slovakian Pinzgau	Holstein
Basic price of milk (SKK/kg)	1.8	1.8	1.8
Price for milk protein (SKK/10g)	1.85	1.85	1.85
Price for milk fat (SKK/10g)	0.9	0.9	0.9
Price of energy in average feed ration for heifers (SKK/MJ NE)	0.727	0.727	0.758
Price of energy in average feed ration for calves (SKK/MJ NE)	0.744	0.744	0.776
Price of energy in average feed ration for cows (SKK/MJ NE)	0.717	0.717	0.748
Price of energy in concentrate for heifers (SKK/MJ NE)	1.154	1.154	1.154
Price of energy in concentrate for cows (SKK/MJ NE)	1.154	1.154	1.154
Price of energy in roughage for heifers (SKK/MJ NE)	0.516	0.516	0.456
Price of energy in roughage for cows (SKK/MJ NE)	0.516	0.516	0.456
Variable labour cost for milk (SKK/kg)	1.1	1.1	0.95

Results and discussion

For considered breeds, the economic weights calculated under milk quota are given in tables 3 and 4. Table 3 shows parameters of milk composites (fat and protein) in %.

Table 3. Economic weights of traits with quota (fat and protein percentage).

	Holstein	Slovakian Simmental	Slovakian Pinzgau
Milk yield (SKK/1kg)	7.17	7.91	8.99
Fat content (SKK/1%)	1828.0	1693.99	1899.08
Protein content (SKK/1%)	11395.05	8996.07	8489.12

Table 4 shows the economic weights for fat and protein yield and amount of milk carier. There were found breed differences in the calculated economic weights. These differences are probably caused by differences in yields and feed energy prices between breeds under analysis. Also, breeds are farmed in different production systems and different circumstances.

Wolfová *et al.* (1996) found differences in the economic weights between Holstein and Czech Pied breeds as well.

Table 4. Economic weights of traits with quota (fat and protein yield in kg).

	Holstein	Slovakian Simmental	Slovakian Pinzgau
Milk yield (SKK/1kg)	0.298	0.510	0.751
Fat yield (SKK/1kg)	32.995	31.356	42.200
Protein yield (SKK/1kg)	158.265	166.593	188.647

The economic weights calculated for production system without milk quota are given in tables 5 and 6. In comparison to calculations with milk quota, the higher economic weights were found for milk yield and fat content as well as amount of milk carier and fat yield. This results from EC regulation on milk and fat. Increasing fat content by 0.1 g per 1 kg of milk reduces milk quota by 0.18%. Similar results were published by Miesenberger (1997), Wunsch & Bergfeld (2001) and Wolfová *et al.* (2001).

Table 5. Economic weights of traits without quota (fat and protein percentage).

	Holstein	Slovakian Simmental	Slovakian Pinzgau
Milk yield (SKK/1kg)	8.038	8.580	9.394
Fat content (SKK/1%)	3033.41	2394.79	2259.84
Protein content (SKK/1%)	11395.05	8996.07	8489.12

Table 6. Economic weights of traits without quota (fat and protein yield).

	Holstein	Slovakian Simmental	Slovakian Pinzgau
Milk yield (SKK/1kg)	1.167	1.183	1.160
Fat yield (SKK/1kg)	42.131	44.348	50.219
Protein yield (SKK/1kg)	158.264	166.594	188.647

On the basis of calculated economic weights, the Slovak Production Index that is a selection criterion applied in Slovakia will be updated. In this index, protein yield will be favoured to fat yield.

References

Adelhelm, R., E. Bőkenhoff, T. Bischoff, D. Fewson & A. Rittler, 1972. Die Leistungsmerkmale beim Rind. Teil A.: Wirtschaftliche Bedeutung der Leistungsmerkmale. Schr.-R. Univers. Hohenheim, R. Tierische Poduction, 64pp.

Groen, A.F., 1989. Economic values in cattle breeding. I. Influences of production circumstances in situation without output limitations. Livest. Prod. Sci., 22: 17-30

Henze, A., J. Zeddies, D. Fewson & E. Niebel, 1980. Nutzen-Kosten-Untersuchungen über Leistungsprüfungen in der tierischen Erzeugung dargestellt am Beispiel der Milchleistungsprüfung beim Rind. Schr. R. Bundesministerium f. Ernährung, Landw. Forsten, R. A. Landwirtschft-Angewandte Wissenschaft, 234.

Miesenberger, J., 1997. Zuchtzieldefinition und Indexselektion für die österreichische Rinderzucht. (Dissertation), Wien, Universität für Bodenkultur.

Sölkner, J., J. Miesenberger, A. Willam, C. Fuerst & R. Baumung, 2000. Total merit indices in dual purpose cattle. Arch. Tierz., 43: 597-608

Wolfová, M., J. Přibyl & J. Wolf, 1992. Stanovení ekonomických vah vlastností mléčné užitkovosti u skotu. Živoč. Výr. 37, 4: 375-381

Wolfová, M. & J. Wolf, 1996. EW Version 1.1 (A PC program for estimating economic weights in cattle). Research Institute of Animal Production, Prague-Uhřiněves, Czech Republic.

Wolfová, M., J. Přibyl & J. Wolf, 2001. Economic weights for production and functional traits of Czech dairy cattle breeds, Czech J. Anim. Sci., 46, 10: 421-432

Wűnsch, U. & U. Bergfield, 2001. Berechnung wirtschaftlicher Gewichte für őkonomisch wichtige Leistungsmerkmale in der Milchrinderzucht. Zűchtungskunde, 73: 3-11

Conclusions

Abele Kuipers

Expertise Centre for Farm Management and Knowledge Transfer, Wageningen University and Research Centre, Hollandseweg 1, 6706 KN Wageningen, The Netherlands
President to be of Cattle Commission of EAAP

In September 2004 the CEEC workshop about farm management, extension needs and milk quota was held in Bled, Slovenia as a one-day satellite workshop of the European Association for Animal Production, EAAP2004. Two preparative workshops were held: in May 2003 a 3-day workshop in Haarlem, The Netherlands and in December 2003 a 2-day workshop in Budapest, Hungary. These meetings helped a great deal in assessing the aspects of importance for the introduction of the quota system and in drawing the scene for developments in animal husbandry in Central and Eastern Europe.

It was very satisfying that in total 107 people registered for the Bled workshop in 2004. Participants from 32 countries were present. Participants came from ministries, research institutes, universities, extension services, farmers' organisations, cattle improvement and business organisations, consultancy services, etc. This mixture of participants from various disciplines including participants from practice was very special. The EAAP always hopes to increase the number of participants from industry and practice. This workshop succeeded in doing so.

All eight new EU member States and seven Eastern European and Balkan countries (Romania, Bulgaria, Byelorussia, Turkey, Croatia, Albania and Georgia) had prepared a country report. Representatives of 13 of these countries presented the report in public at the meeting in the afternoon session. One contribution summarised farm management efforts under quota conditions in the time to come. In the morning session, specialists and a dairy farmer from countries with long time history with quota told about the developments in the dairy sector and market and about their experiences with the quota system.

All contributions were very well prepared. Everybody was present, so there was not one no-show! Two presentations were duo-presentations. For instance, the presentation from Slovakia was from the Director of the Slovak Association of Milk Producers together with a researcher from the Animal Production Institute. It appeared that a one-day workshop with so many participants appeared to be too short. Discussion is also a very important part of a workshop. This requires more time.

Some impressions from the workshop are:
- The effects of the quota system on dairy husbandry is not yet fully realised in the new EU countries. This is caused by the fact that the national milk quota is presently not exceeded by the total production of the farmers resulting in less pressure to adapt the production volume. This will probably change in years ahead.
- Administrative procedures concerning the quota system still receive a lot of attention by the central authorities in the new countries. Extension efforts concerning aspects of farm management, farm strategy and sector development are in an initial stage.
- This workshop increased the awareness of the participants towards questions which may be expected from practice.
- The candidate EU countries demonstrated a keen interest in the modalities of setting -up a quota system in their countries. They are in the stage of designing the base of their system.
- Exchange of information and experiences between countries appeared to be very helpful.

Appendix I

Summary of 3-day workshop in the Netherlands: Introduction aspects of milk quota systems in EU candidate countries in Harlem, The Netherlands, May 26, 27 and 28th 2003

Eastern and Central European countries entering the EU in 2004 should have implemented the milk quota system by 1st April 2004. The introduction of the milk quota system is a laborious task. All dairy farmers need to be assigned an individual quota. But also choices have to be made about:

*   free transfer of quota versus a quota bank;
*   quota linked to land or not;
*   creation of a national reserve;
*   selection of priority groups of farmers;
*   using quota as a tool for regional development and differences or not;
*   leasing of quota or not;
*   etc.

Technical aspects, such as the butterfat correction ask for detailed calculations. For direct sales from the farm, a different set of rules has been used as for dairy deliveries. The quota system requires a control apparatus. Moreover, the management of the farm under a quota system needs adaptation. Introduction of quota also have impact on other agricultural sectors and the industry. Dairy farmers tend to diversity their farm operation. Some farmers leave the dairy sector. Inputs from industry change.

A 3-day seminar was organized to discus all these aspects that accompany the introduction of a quota system. The seminar was held in The Netherlands. The time that elapsed between the initial idea of a seminar and the actual seminar was less than two months. All participating parties did a great job in preparing the seminar in such a short time.

Ten Eastern and Central European countries send a delegation. Estonia, Latvia, Lithuania, Czech Republic, Hungary, Slovakia and Slovenia were present as candidate countries entering the EC in 2004. Only Poland had to withdraw its delegation at the last moment, because of urgent work to be done at home on the introduction of quota. Romania, Bulgaria en Turkey participated as countries probably entering the EU in a later stage.

During the seminar, experiences with the introduction of quota in Western-European countries were presented. All introduction aspects were discussed in more or less detail. The application of certain aspects was also considered taking into account the situation in Eastern European countries.

The starting positions of the 8 candidate countries entering in 2004 differ widely. The participants reported on the present state of introduction of quota in their respective countries. Data on the countries were obtained through a questionnaire as well.

The number of farmers with dairy cows varies from more than 200.000 (Lithuania) to 3.500 in the Czech Republic. But in Lithuania, 80% of farms have 1 or 2 cows, while in the Czech Republic the average number of cows per farm is 212. Another country with large herds is Hungary (42% of farms > 100 cows). In all other countries, more than 80% of the farmers have 10 or less cows. Many of these small farmers will use milk and milk products for home consumption. They do not deliver milk to others (to purchasers). It may be practically impossible to register all of these producers. It is also practically impossible to control this group conform EU regulation. For the direct-sales categories, it is very important to have simple and applicable rules.

Most countries are still working on a national quota regulation, while some have still to assign individual quota to the farmers. In Slovenia, the start of the official implementation of quota is postponed to 1st April 2005. Most counties have still to decide about important aspects as transfer of quota, leasing of quota, etc.

In Hungary, a production supply system was already in operation for several years. But the butterfat reference and some other EU rules are also new elements to be implemented in this country. In Poland, the introduction of a quota system is already some years in preparation. But in almost all countries, a lot of work has still to be done and decisions to be made.

The estimated number of expected recognized purchasers varies from 86 in Latvia to 20 in the Czech Republic. The starting position in the various countries also differs, because of large variations in production pressure. In some countries, milk volume is far below the assigned national quota amount because of negative developments in the dairy sector. On the opposite side, farmers in some other countries exceed already their national quota. In these countries, the introduction phase of the quota system will be more difficult for the farmers than in countries with a milk volume below the national quota.

Estimated direct sales differ from 1% of total milk production in the Czech Republic and 6% in Estonia to 24% in Lithuania, 25% in Slovenia and 33% in Latvia.

Some of the interesting observations made during the seminar are listed below:
- Not every farmer requests a quota immediately. Request from farmers will come much later as well.
- Purchasers should be recognized as early as possible to avoid problems later on.
- Preferential treatment of farmers in assigning quota causes easily problems. If possible, limit the number of categories that will receive a special treatment.
- A system with (high) quota prices results in an increase of fixed costs on farm level. This is especially of concern to future generations of farmers. With free transfer of quota, professional middlemen bring buyers en sellers of quota together. They act as paid intermediaries.
- A quota bank collects the free quota from farmers and transfers it to other farms, who demand additional quota. In this way, transfer of quota is regulated centrally. It also provides the opportunity to give a preferential treatment to some categories of farms / farmers. This can be done to help "hard cases" or to stimulate certain structural developments in the sector, or for other purposes.
- The instrument of leasing benefits farmers, who retire from the dairy farm. In this case, non-farmers rent quota to the active farmers. This slows down somewhat the structural developments in the dairy sector.
- The fat reference method gives many questions concerning the implementation and calculation
- For direct sales, there are two main methods for establishing a quota per farm: a. number of cows per farm x average production of region or b. product equivalents as base.
- For direct sales, there is a tendency to advocate a more simple approach.
- Within the national quota system exists the possibility to establish quota per region. This possibility seems to be used only rarely.
- One unique registration number for all relations to be used for all regulations is very useful.
- Depending on the number of farms and control items, it is recommended to use scan equipment and software for registering the forms.
- When applying quota, a farmer needs to review its strategy for the farm (its future goals) and to adapt his management to the quota restrictions. Other priorities may be chosen.

- In general, the quota system will result in the long run in shifting part of the labor to other agricultural activities and to non-agricultural employment.
- Communication around the introduction of quota is often neglected. That is a mistake. It is advised to look for partners who can assist in performing this important task.

Appendix II

Minutes of 2-day workshop in Budapest: Agricultural development, dairy farm management and extension needs in CEE countries under the restrictions of the EU milk quota in Budapest, Hungary on 1st and 2nd of December 2003

Workshop "Agricultural developments, dairy farm management and extension needs in CEE countries under the restrictions of the EU milk quota" was held on 1-2 of December 2003 in FAO Sub-regional Office for Central and East Europe, Budapest.

Present: 14 participants from FAO, Lithuania, Czech Republic, Poland, Slovakia, Hungary, Slovenia, Romania, Germany and The Netherlands.

1. Opening of the meeting by the Project Co-ordinator and the approval of the Agenda

Sub-regional Representative for Central and Eastern Europe, Maria Kadlecikova opened the meeting and welcomed all in attendance.

Abele Kuipers, Vice-President of the Cattle Commission of the EAAP and Arunas Svitojus, Chairman of the CEE Working Group presented an overview of work that is being done and in preparation about milk quota and premium systems and it's consequences on agriculture as part of the introduction of the Central European States into the European Union.

a. Milk quota seminar in The Netherlands, May 2003

In May 2003, a milk quota seminar was held in The Netherlands. It concentrated on the administrative aspects of the introduction of the quota system. About 30 participants from the Central European countries attended this 2 ½ day workshop. A report was published. A few of the participants here in Budapest also visited the meeting in The Netherlands.

b. Workshop in Budapest, December 2003

The current workshop in December 2003 in Budapest about the milk quota system (see for objectives also point 2 of this Summary Minutes) concentrates on the consequences of the quota and premium system for agriculture and is a preparatory meeting for the Seminar in Slovenia in September 2004.

c. Seminar in Bled, Slovenia, September 2004

The seminar in Bled, Slovenia on Saturday the 4th of September 2004 as satellite symposium of the EAAP (European Association for Animal Production) meeting has as title: "Farm management and extension needs in CEE countries under the restrictions of the EU milk quota". During the EAAP meeting, there is also a Round Table (Title: "Enlargement of the European Union and Other Challenges for the European Livestock Production") discussion about the enlargement of the EU with 10 new member States in 2004.

The program of the Bled seminar will be discussed at the second day of this workshop in Budapest. The program will be adjusted using the experiences we gain during this workshop. Also the agenda for this Budapest workshop was presented. The agenda was approved as proposed.

2. Objectives of the workshop:

- Exchange of information on latest developments in agriculture in EU accessing and some other CEE countries
- Review of farm management adaptation to milk quotas and premiums systems and related structural changes in the dairy and other agricultural sectors
- Organisation and preparation of the EAAP Symposium in Bled, Slovenia, 2004; Search for a co-operative project concerning the topics to be discussed in workshop.

3. Welcoming address and a short presentation of the Host Partner

Maria Kadlecikova presented a topic about developments in the Central and East European countries in the view of the EU accession. She mentioned the importance of preparing in an early stage for the consequences of the EU legislation, such as milk quota.

Abele Kuipers made a presentation about the description of different milk quota systems in the EU Member states. The main elements of this system were explained. Also some applications of the quota system which will most probably effect the structure of dairy husbandry in the CEE countries were addressed.

After the presentation, Mr. Kuipers offers Mrs. Kadlecikova the report of the quota seminar in The Netherlands.

Mr. Heinrich Hockmann his topic was about structural developments as consequences of the introduction of milk quota and suckler cow and beef premiums, principles and theory of analysing and predicting structural developments in animal husbandry, milk processing and marketing.

Mr. Stjepan Tanic discussed about sustainable rural livelihood, farm income and production diversification policies in context of the EU accession.

Mr. Clemens Fuchs presented the effects on farm management and also extension needs due to quota system and premiums EU data from some Central European countries as well as field experiences from Eastern Germany were part of this presentation.

All participants from the 7 Central European countries presented 1 till 3 country reports about:

- Structural changes to milk quota and premium system.
- Effect of quota on farm management.
- Extension needs

The participants from the various countries presented an inventory of studies, experiences and personal thoughts about the developments in agriculture due to introduction of milk quota and premiums. The country reports were well prepared.

A discussion was held about the issues of structural developments, farm management under milk quota regimen and extension needs in the Central European countries. Interesting views were expressed and information provided.

4. Discussions about satellite symposium, which will be organized on 4th of September 2004 at 55th Annual Meeting of the EAAP

Discussions on the program for EAAP symposium in Bled: topics, timetable, and speakers to be invited.

The discussion concentrated on which CEE countries to invite to this meeting and how to deal with the country reports. It is difficult looking at the available time to have 15 or more presentations of country reports in one day of seminar. Otherwise, information from these countries is very much needed as input for the seminar. Perhaps we can summarise the country reports in one overview paper and presentation.

We will formulate a questionnaire as basis for the country reports to be presented during the meeting in Bled: some questions for the 8 countries in the region joining the EU in 2004 and somewhat different questions for other CEE countries. The last mentioned countries can not refer to the current introduction of the quota system in their country.

5. Any other business
- o Implementation of new projects according to milk quota between different countries.
- o Discussions about support from FAO and TAIEX for EAAP quota symposium in Bled.

6. Closing of the meeting
Arunas Svitojus, Chairman of the CEE Working Group and Abele Kuipers, Vice-President of the Cattle Commission of the EAAP thanked all delegations for their participation at "Agricultural development, dairy farm management and extension needs in CEE countries under the restrictions of the EU milk quota" workshop. Especially, appreciation was expressed for the representatives of the FAO/SEUR office for the help in organizing this workshop and the pleasant welcome and good opportunity to meet colleagues. Mr. Leos Celeda was the contact point at the Budapest office and deserves credit for his efforts and help. The FAO also sponsored this workshop.

The EAAP Technical Series so far contains the following publications:

- **No. 1. Protein feed for animal production**
  With special reference to Central and Eastern Europe
  edited by C. Février, A. Aumaitre, F. Habe, T. Vares and M. Zjalic
  ISBN-10: 90-76998-03-5 – ISBN-13: 978-90-76998-03-9 – 2001 – 184 pages – € 35 – US$ 39

- **No. 2. Livestock breeding and service organisations**
  With special reference to CEE countries
  edited by J. Boyazoglu, J. Hodges, M. Zjalic and P. Rafai
  ISBN-10: 90-76998-04-3 – ISBN-13: 978-90-76998-04-6 – 2002 – 75 pages – € 25 – US$ 30

- **No. 3. Livestock Farming Systems in Central and Eastern Europe**
  edited by A. Gibon and S. Mihina
  ISBN-10: 90-76998-29-9 – ISBN-13: 978-90-76998-29-9 – 2003 – 264 pages – € 39 – US$ 51

- **No. 4. Image of the Cattle Sector and its Products**
  Role of Breeders Association
  edited by J. Boyazoglu
  ISBN-10: 90-76998-33-7 – ISBN-13: 978-90-76998-33-6 – 2003 – 88 pages – € 27 – US$ 32

- **No. 5. Foot and Mouth Disease**
  New values, innovative research agenda's and policies
  A.J. van der Zijpp, M.J.E. Braker, C.H.A.M. Eilers, H. Kieft, T.A. Vogelzang and S.J. Oosting
  ISBN-10: 90-76998-27-2 – ISBN-13: 978-90-76998-27-5 – 2004 – 80 pages – € 25 – US$ 30

- **No. 6. Working animals in agriculture and transport**
  A collection of some current research and development observations
  edited by R.A. Pearson, P. Lhoste, M. Saastamoinen and W. Martin-Rosset
  ISBN-10: 90-76998-25-6 – ISBN-13: 978-90-76998-25-1 – 2003 – 208 pages – € 40 – US$ 53

- **No. 7. Interactions between climate and animal production**
  edited by N. Lacetera, U. Bernabucci, H.H. Khalifa, B. Ronchi and A. Nardone
  ISBN-10: 90-76998-26-4 – ISBN-13: 978-90-76998-26-8 – 2003 – 128 pages – € 35 – US$ 39

These publications are available at:
Wageningen Academic Publishers
P.O. Box 220
6700 AE Wageningen
The Netherlands
sales@WageningenAcademic.com
www.WageningenAcademic.com

Wageningen Academic
P u b l i s h e r s

Printed in the United States
by Baker & Taylor Publisher Services